Shadows over Sunnyside

An Arkansas Plantation
in Transition, 1830–1945

EDITED BY
Jeannie M. Whayne

THE UNIVERSITY OF ARKANSAS PRESS
FAYETTEVILLE   1993

97   96   95   94   93   5   4   3   2   1

Designed by Gail Carter

The paper used in this publication meets the minimum requirements of the
American National Standard for Permanence of Paper for Printed Library Materials
Z39.48-1984. ∞

Library of Congress Cataloging-in-Publication Data

Shadows over Sunnyside : an Arkansas plantation in transition, 1830–1945 / edited by
    Jeannie M. Whayne.
        p.   cm.
    Includes bibliographical references and index.
    ISBN 1-55728-283-8 (c)
        1. Sunnyside Plantation (Ark.)—History.   2. Italian Americans—Arkansas—
Chicot County—Economic conditions.   3. Peonage—Arkansas—Chicot County—
History.   4. Cotton growing—Arkansas—Chicot County—History.   5. Chicot
County (Ark.)—History.   6. Chicot County (Ark.)—Economic conditions.
I. Whayne, Jeannie M.
F419.S96S48   1993
976.7' 84—dc20   92-39706
                                                                      CIP

Contents

Contributors

RANDOLPH H. BOEHM is a senior editor of University Publications of America, Bethesda, Maryland. He is currently at work on a full-length study of Mary Grace Quackenbos.

PETE DANIEL is curator of the Division of Agriculture and Natural Resources, National Museum of American History, Washington, D.C., and an authority on twentieth-century peonage in the South. He has published several books on the twentieth-century South.

WILLARD B. GATEWOOD is Alumni Distinguished Professor of History at the University of Arkansas, Fayetteville. He is currently at work on a full-length study of Elisha Worthington's Sunnyside Plantation. He is author of several books and numerous articles.

ERNESTO R. MILANI received a degree in modern languages from the University of Milan with a thesis on mutual-aid societies among Italian immigrants in the United States. His current project is a microanalysis of Italian settlements in the United States and Canada.

GEORGE E. POZZETTA is a professor of history at the University of Florida, Gainesville. His research, writing, and teaching have centered on the history of American immigration and ethnicity, and he is the author of several books and essays on immigration history.

JEANNIE M. WHAYNE is an assistant professor of history at the University of Arkansas, Fayetteville, and editor of the *Arkansas Historical Quarterly*.

She is currently engaged in a full-length study of the evolution of a plantation system in the Arkansas Delta from 1900 through 1970.

BERTRAM WYATT-BROWN is the Richard J. Milbauer Professor of History at the University of Florida, Gainesville. He is the author of a comprehensive study of the Percy family.

Preface

This book originated at a session of the 1990 meeting of the Organization of American Historians. That session focused on an experiment with Italian immigrant labor at Sunnyside Plantation in the late nineteenth and early twentieth centuries. Ernesto Milani told the story from the Italian perspective; Randolph Boehm focused on the role that federal investigator Mary Grace Quackenbos played; Bertram Wyatt-Brown examined Leroy Percy's involvement; and Pete Daniel wrote the commentary. The three papers and the commentary were then published in the Spring 1991 issue of the *Arkansas Historical Quarterly* along with an additional article written by Willard B. Gatewood. Gatewood placed the Italian experiment into the context of the evolution of the Sunnyside Plantation from its antebellum origins to its use as a resettlement community in the 1930s.

This volume reprints three of the four articles largely as they appeared in the Spring 1991 issue of the *Arkansas Historical Quarterly*. Willard Gatewood, who has been engaged in a book-length study of the Sunnyside Plantation, has made certain revisions to his original article.

The current book is divided into two parts. Gatewood's overview of Sunnyside begins Part I and is joined by a historiographical article written by Jeannie Whayne, who places Sunnyside within the context of the historical literature on the evolution of the southern plantation following the Civil War. Part II consists of the essays by Milani, Boehm, and Wyatt-Brown followed by a historiographical analysis written by George Pozzetta, who places the use of Italian immigrant labor into the context of the scholarly literature on the mass migration of Italians in the late nineteenth and early twentieth centuries. Daniel's commentary appears in the Appendix along with several documents designed to highlight points made by the various authors.

It is the hope of the Arkansas Historical Association that readers will find this volume useful in understanding the historical significance of the history of the Arkansas Delta. Too often ignored by historians of the South, Arkansas and its Delta deserve to be comprehended as a part of the southern experience. We believe that this work demonstrates that the issues affecting the rest of the South before and after the Civil War were of equal significance in Arkansas and that an understanding of Arkansas's place in this drama helps to illuminate the southern experience as a whole.

The authors and editor wish to thank a number of individuals and institutions for their assistance. David Moyers of Hamburg, Arkansas, and Ms. Kathy P. Johnson, County and Probate Clerk of Chicot County, Lake Village, Arkansas, graciously responded to Gatewood's frequent requests for aid. The National Endowment for the Humanities, the Earhart Foundation, and the National Humanities Center provided generous support for Wyatt-Brown's, who also wishes to acknowledge the assistance of Randolph Boehm, Pete Daniel, Stanley Engerman, Richard Bernstein, and the members of the North Carolina Economic Historians Seminar for their criticisms and suggestions. The Mississippi Department of Archives and History responded promptly to the editor's request for photographs relating to Sunnyside. The editor is especially grateful to Randolph Boehm for his tireless work in locating photographs and documents for the appendix.

Shadows over Sunnyside

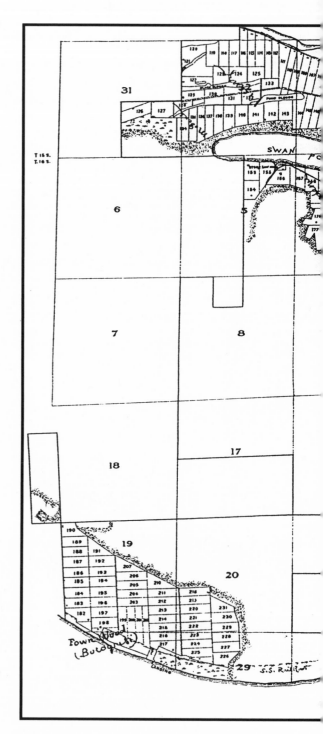

Map of Sunnyside
showing regional
settlements by
Father Bandini.
(C.S.E.R.—Rowe—
Private Collection of
Ernesto R. Milani.)

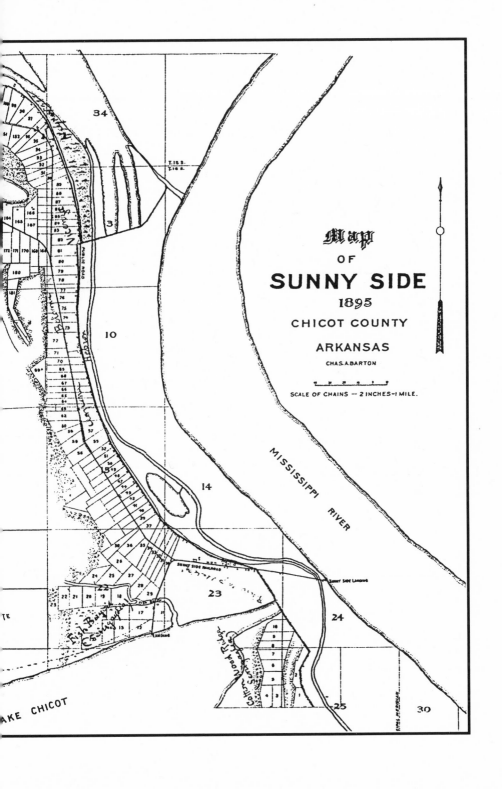

Map

OF

SUNNY SIDE

1895

CHICOT COUNTY

ARKANSAS

CHAS. A. BARTON

SCALE OF CHAINS — 2 INCHES—1 MILE.

MISSISSIPPI RIVER

LAKE CHICOT

PART ONE

Sunnyside: The Evolution of an Arkansas Plantation, 1840–1945

Willard B. Gatewood

Sunnyside Plantation in Chicot County, Arkansas, existed more or less intact for over a century after its creation in the 1830s. Originally a single plantation whose resident owner possessed numerous slaves, it later became the centerpiece of a series of plantations owned by successive absentee corporations which, except for an experiment with Italian farmers, usually employed black labor in some form of sharecropping. Spanning the period from the westward movement of cotton in the early nineteenth century through the Civil War and Reconstruction to the New Deal and World War II, the history of Sunnyside offers revealing glimpses into the important events, forces, and developments that shaped cotton agriculture in the lower South for more than a hundred years. In the early twentieth century, Sunnyside achieved brief notoriety because of charges of Italian peonage.

Located in the southeastern corner of Arkansas on the Mississippi River, Sunnyside Plantation in 1860 consisted of three thousand acres of incredibly fertile alluvial land. It possessed easy access by water to New Orleans, and the topographical peculiarity of its location between the river and Old River (or Chicot) Lake, the former channel of the Mississippi, meant that Sunnyside was less susceptible to floods or "overflows" than many other plantations in the vicinity. Beginning early in the 1830s, when Arkansas was a sparsely settled territory, the Sunnyside Plantation's cultivated acreage steadily increased as more land was cleared of dense forest, thickets, and canebrakes to make way for an ever-expanding cotton culture. As Sunnyside and neighboring plantations grew in size and complexity, they came to resemble the more mature plantations in older portions of the South. Family and family connections figured significantly in the life of antebellum Chicot County. Deeds, wills, and other official records provide abundant evidence of the extent to which a network of families, small in number and often related by blood or marriage, perpetuated their dominant economic and social position. By the end of the antebellum era, when Chicot County had become one of the wealthiest of the so-called cotton counties in Arkansas, its largest and most prosperous plantation was Sunnyside.[1]

The founder of Sunnyside Plantation was Abner Johnson, a native of Kentucky, who migrated in the 1820s to the wilderness of southeast Arkansas to settle in what became Chicot County. Into this wilderness came settlers from Kentucky, Mississippi, Tennessee, and other southern states in search of cheap land. "It took a stout heart [and] much bravery," a local historian remarked, "to enter and establish a claim in this dense-wooded wilderness."[2] Abner Johnson obviously possessed such qualities. Elected sheriff in 1832, he continued throughout the 1830s to acquire numerous tracts of fertile land at low prices.[3] By 1840, however, Johnson and his wife had decided to dispose of their land and return to Kentucky.

Before his departure Johnson sold his largest single tract of land, known as Sunnyside Plantation, to Elisha Worthington, another Kentuckian who had settled in Chicot County a few years earlier. The sale in 1840 involved 2,215 acres of land, 42 slaves, and a variety of buildings at a total cost of $60,000. The mortgage required Worthington to remit to Johnson 250 bales of cotton annually for ten years.[4] Except for two relatively small parcels of land, Worthington did not acquire additional acreage until after he had paid his debt to Johnson. But during the decade of the 1840s, he borrowed against Sunnyside, executing second mortgages, to inaugurate improvements on the plantation and to bring more of its undeveloped land under cultivation.[5]

Elisha Worthington, who actually transformed Sunnyside into a highly productive plantation with its own "landing" on the Mississippi, was one of three brothers of an old and respected Kentucky family that migrated south and settled in the Mississippi Valley. Two of the brothers settled across the river from Chicot County in the vicinity of Greenville, Mississippi. Elisha Worthington's nephew, Isaac, moved to Chicot County after the Civil War and married Mary Johnson, whose family was considered by some as the "most aristocratic" in Chicot

County. Mary Johnson was the daughter of Lycurgus Johnson, the owner of Lakeport Plantation, whose family was politically influential in antebellum Arkansas.[6] Although Isaac Worthington, his wife, and her family were on cordial terms with his uncle Elisha, they probably disapproved of his lifestyle. Like Mary Worthington's illustrious kinsman, Vice President Richard M. Johnson, who had a family "across the color line,"[7] her husband's uncle openly violated the prevailing racial code in Chicot County.

Elisha Worthington married a young woman in Kentucky on November 10, 1840. He brought his new bride to Chicot County to live at Sunnyside Plantation, but less than six months later she returned to Kentucky because "of the adultery of her husband." By a special act in 1843 the Kentucky legislature annulled "the bonds of matrimony between her and her husband."[8] Worthington never remarried but appears to have maintained a stable relationship with a black woman, a slave. The identity of his consort is not known, but her name was probably Mason since the two children born to her and Worthington were James W. and Martha W. Mason. Worthington not only was an indulgent father and reared these children at Sunnyside, but he also educated them. Both attended Oberlin in Ohio, and the son studied in France until 1860 when he returned to Sunnyside. That Worthington maintained a biracial household and recognized his mulatto children may well account, at least in part, for the slight notice that he has received in the historical record of the county. Local chroniclers have lavished attention on such cotton "aristocrats" as the Johnsons, Hilliards, Gaineses, and a few others, but have referred to Elisha Worthington primarily as a large landowner who "died without heirs," despite the fact that his mulatto daughter waged a protracted and ultimately successful battle to share in his estate.[9]

By the time his son returned to Sunnyside from France in 1860, Worthington commanded great respect in spite of his living arrangements. The sheer size of his wealth made him one of the most influential planters in southeast Arkansas. Invariably referred to as "Colonel" (for reasons that are not clear), he was the largest land- and slaveowner in a county where slaves outnumbered whites by a margin of more than three to one. He traveled extensively and had important business dealings with individuals and commercial houses in Louisiana, Mississippi, and Kentucky, as well as in Arkansas.

During the decade of the 1850s, Worthington substantially expanded his cotton lands by purchasing fifteen additional tracts ranging in size from 39 acres to 1280 acres, all located on or near Lake Chicot. The acquisition of such prime cotton land meant that throughout the 1850s he borrowed heavily and regularly. He mortgaged lands free of encumbrances to secure loans to purchase additional tracts. About 800 acres of this new land became incorporated into Sunnyside Plantation, which by 1860 had 1,300 acres under cultivation and 1,700 acres in virgin forests of cottonwood, gum, and oak. Sunnyside, which was debt free by the early 1850s, served as collateral for some of Worthington's loans later in the decade.[10]

The largest single debt incurred by Worthington was for $274,000 in 1858. The impact of the financial crisis of 1857 may partially explain his need for so large an amount, but he apparently used a portion of it to consolidate and pay off other loans and to purchase more land. This loan was made by Abraham Van Buren and his wife, Angelica Singleton Van Buren, and Wade Hampton of South Carolina, whose large plantation "Wild Woods" lay across the river near Greenville, Mississippi. Worthington and Hampton undoubtedly were acquaintances in view of the proximity of their plantations. Just how the Van Burens came to figure in this loan is unclear. Abraham Van Buren, the son of President Martin Van Buren, was a New York financier whose wife served for a time as White House hostess for her father-in-law. The fact that the Singletons and the Hamptons were related may account for the Van Buren participation in the loan.[11]

By 1860 Elisha Worthington owned 543 slaves and approximately 12,000 acres of choice land in Chicot County. He was the largest slaveowner in Arkansas. His taxable property was valued at $472,000, more than twice that of any other planter in the county. Between 1850 and 1860 Worthington quadrupled the amount of land he had under cultivation from 800 acres to 3,600 acres. Sunnyside, where he resided, remained the centerpiece of his ever-expanding domain. In addition, he owned three other plantations: Redleaf, which rivaled Sunnyside in size and cotton production, Meanie, and Eminence, which contained far less cultivated acreage. Each plantation had a "residence and quarters," and Redleaf possessed a steam cotton gin and grist mill. But Sunnyside was home for almost half of the Worthington slaves, and in 1860 it produced 1,700 bales of cotton, while Redleaf with about 120 slaves produced 1,270 bales. Cotton was not cultivated on the other two plantations, which concentrated on raising corn, sweet potatoes, pigs, and cattle. Altogether the Worthington plantations in 1860 contained 16 horses, 200 mules, 80 working oxen, 45 milk cows, 650 head of cattle, and 340 pigs. In addition to 2,950 bales of cotton, these plantations produced 31,500 bushels of corn, 2,200 bushels of sweet potatoes, and 1,000 pounds of butter, as well as poultry, beeswax, Irish potatoes, vegetables, and fruits.[12] Such diverse productivity suggests that Worthington's plantations had achieved a substantial degree of self-sufficiency by 1860.

The wealth derived from the fertile alluvial lands of Chicot County spawned a proliferation of plantations, such as Florence, Patria, Pastoria, Luna, and Lakeport,[13] but none was so large or as productive as Worthington's Sunnyside. Like other planters, Worthington was profoundly affected by fluctuations in the cotton market. During his ownership of Sunnyside (exclusive of the Civil War years), the price of cotton on the New Orleans market ranged from an average low of 5.5 cents per pound in 1844 to an average high of 12.4 cents in 1856.[14] The high prices in the last half of the decade of the 1850s brought great prosperity to Sunnyside and other Chicot County plantations. During these years, as a frequent northern visitor later recalled, Chicot was "the richest, fairest and most productive county in the state." Most impressive of all, in his view, were those plantations such as Sunnyside located on Lake Chicot and "entirely above overflow which . . . were like a continuous garden, all under cultivation, raising a bale of cotton to the acre, with elegant houses, negro quarters, stables, etc."[15] Notwithstanding the exaggeration born of the writer's nostalgia for a scene destroyed by the Civil War, his description of the cotton fiefdoms in Chicot County obviously contained some truth.

The cotton plantations of Chicot County, which had constituted "as fair a picture as one could wish to see,"[16] bore the brunt of the chaos and destruction of the Civil War, especially during its final two years. Late in 1862, when it became apparent that Chicot County would become a battleground in the war, Worthington took most of his slaves and livestock to Texas, leaving his son and daughter, James and Martha Mason, in charge of Sunnyside Plantation. Because of its strategic location on the Mississippi River, the plantation was all the more vulnerable to plunder and devastation by Confederate and Union forces. To thwart further Confederate success in disrupting river traffic, the Union army prepared to "clean out the rebels" in the area by sending a large force to Sunnyside Landing on June 5, 1864. Following a skirmish between Union and Confederate troops at Redleaf, a major clash occurred on June 6 at the Battle of Ditch Bayou, which a recent historian has described as a costly encounter "fought by a transient Union force to destroy or disperse a Confederate force no longer capable of disrupting river traffic effectively and which was on the verge of being ordered elsewhere."[17]

Late in 1865 Elisha Worthington, then in his mid-fifties and a virtual invalid, returned to Chicot County from Texas with most of those who had previously been his slaves. In addition to the bloodshed of battle, the military action in Chicot County obviously had a devastating impact on Worthington's plantations, especially Sunnyside and Redleaf. Worthington faced an uncertain future: slavery had been abolished, and the armies that traversed his lands had left disorder and destruction. Whether or not he was aware of it, the Civil War and the end of slavery coincided with another major development certain to affect his future—"the close of a long period of rapid growth in world demand for cotton."[18] Of immediate concern, however, was the fact that his plantations, classified as "abandoned lands," were under the control of the Freedmen's Bureau. Weeds and bushes had overtaken fields that once produced cotton, corn, and other crops. Pardoned by President Andrew Johnson on January 31, 1866, Worthington regained his four plantations and attempted to rehabilitate them. Employing principally black workers, including many of his former slaves, he acquiesced in the free labor system supervised by the Freedmen's Bureau. This system at first involved only wage labor, but quickly came to embrace what was known as "the share of the crop," or sharecropping.[19]

Because of ill health and mounting financial problems,[20] Worthington began to dispose of his property in the spring of 1866. Through a complicated arrangement with Robert P. Pepper, an old acquaintance in Kentucky, Worthington disposed of 3,895 acres of his land, all of which he bought back "except so much as was embraced in the Sunny-Side Plantation."[21] Therefore, his prize plantation, or a large portion of it, passed into the hands of the Pepper family. The proceeds of this sale enabled him to pay off his debt to Wade Hampton and the Van Burens in 1867. After selling Sunnyside, Worthington moved to Redleaf, where he and his daughter, Martha, who served as his housekeeper, nurse, and business agent, lived for the next seven years. When Worthington died in 1873 without a will, his estate became the subject of protracted litigation by various creditors and his daughter.[22]

In the meantime, in 1868, the William Starling Company, a family-owned enterprise in Chicot County, purchased Sunnyside Plantation from the estate of Robert Pepper, whose family was related by marriage to the Starlings.[23] The purchase of the plantation by the Starling Company coincided with the beginning of a turbulent era of Reconstruction in Chicot County. The local Freedmen's Bureau agent received so many complaints about the Starlings' disregard for Bureau regulations in their management of Sunnyside that he finally threatened to remove all freedmen from the plantation and to send to Vicksburg for troops "to make the necessary changes."[24]

In the meantime, Worthington's son, James W. Mason, had emerged as the Republican "boss" of the county. A state senator and later county sheriff, Mason was accused of inspiring the so-called "Chicot Massacre" of 1871, a violent racial clash that prompted the flight of many whites from a county whose population was overwhelmingly black.[25] The return of the Democrats to power in Arkansas in 1874 and the death of Mason later the same year meant that white planters once again resumed a substantial measure of control over Chicot County affairs, even though blacks continued to monopolize local offices for another decade and a half.[26] Despite difficulties with the Freedmen's Bureau and the turbulence in the county during the early 1870s, the Starlings held on to Sunnyside Plantation, which apparently achieved a degree of its former prosperity by the middle of the decade.

A marriage that took place on December 8, 1870, between an heir to a Chicot County plantation and a descendent of one of the South's most eloquent antebellum spokesmen was destined to add another chapter to the complicated history of Sunnyside Plantation: John C. Calhoun, the grandson and namesake of the famous South Carolinian, married Lennie Adams of Kentucky, whose mother was Betsy Johnson from Chicot County. Through her mother, Lennie Adams inherited the 2,264-acre Florence Plantation. Prior to their marriage, Calhoun, a Civil War veteran, had organized a partnership with James R. Powell of Montgomery, Alabama, in 1866 "for the purpose of colonizing negroes in the Yazoo Valley, Miss., to work plantation lands on the cooperative basis." This enterprise moved blacks from the poorer or worn-out cotton lands of the southeast into the newer plantations of the rich Mississippi alluvial plain where there was a labor shortage. The Calhoun-Powell venture proved to be so profitable that Calhoun sold his half-interest for $10,000 after one year and headed for Arkansas to "repeat the enterprise on a much larger scale." In 1869 he settled in Chicot County and was joined there briefly by his brother, Patrick, an attorney. Although Patrick Calhoun moved to Atlanta where he became a wealthy and influential corporation lawyer and entrepreneur, he remained intimately involved in his brother's land projects in Arkansas.[27]

John C. Calhoun probably first met Lennie Adams in Chicot County, where she often visited her aunt and guardian, Julia Johnson Johnson, the widow of the former governor of Louisiana, who resided at her family's

Lakeport Plantation. As of the census of 1870, Calhoun owned no land in the county, and for the next eight or nine years he appears to have focused his energies on the cultivation of cotton on his wife's Florence Plantation to which she retained legal ownership and designated him as manager.[28] The plantation prospered, enabling Calhoun and his wife to borrow money for purchasing and improving more land. In time Calhoun became a highly visible planter in southeast Arkansas. In 1883 the state legislature appointed him a levee inspector. He represented Arkansas at the cotton expositions in Louisville and New Orleans in 1883 and 1884 respectively, and he served in the latter year as vice president of a convention in Washington that petitioned Congress for levees and other improvements on the Mississippi River and its tributaries.[29]

In 1881, Calhoun, his wife, and his brother Patrick began securing options on large tracts of land in Chicot County. By mid-1882 they had acquired options on some of the best-known plantations in the county, including Harwood, Hebron, Luna, Fawnwood, Patria, Hyner's, and Latrobe, which they ultimately purchased. Their most expensive acquisition was Sunnyside Plantation, bought from the Starling family for $90,000. Because Julia Johnson was so heavily in debt to her niece, Lennie and John Calhoun also took "charge and control" of her Lakeport Plantation.[30]

To finance the purchase of these plantations, the Calhouns organized the Calhoun Land Company, a stock company incorporated in Connecticut in 1881. This company was followed by two others, the Florence Planting Company and the Chicot Planting Company, whose complicated relationship must await the careful scrutiny of a more knowledgeable historian.[31] Suffice it to say that these enterprises had the backing of eastern financiers, including J. Baxter Upham of Boston and Austin Corbin, a banker, railroadman, and developer in New York. In fact, Corbin's bank held mortgages on most of the Calhoun's property and acted as the trustee for the bonds issued by their companies. One feature of the Calhoun enterprise that undoubtedly made it attractive to eastern capitalists was that the lands already acquired as well as those the company planned to purchase were located in the vicinity of projected routes of several railroads.[32]

While Patrick Calhoun served as the behind-the-scenes organizer and negotiator for the family's land enterprises, his brother John C. Calhoun articulated their aims and ambitions. In discussing his land company with the press early in 1882, he roundly condemned the old plantation system that made planters "totally dependent and helpless" in the face of forces beyond their control and that "debased the negro masses." The way out of this predicament, he argued, was for agriculture to undergo consolidation and crop diversification, to introduce business methods, and to abandon the practice of planters having to be "carried through every season, over-burdened with obligations." As earlier, when he interpreted his colonization of blacks in the Mississippi Valley as a means of contributing to the economic advancement of the newly freed slaves, he spoke of his endeavors in Arkansas in a similar manner. Calhoun announced plans for acquiring 20,000 additional acres of land "on the opposite bank" of the river in Mississippi to add to his Arkansas domain. It was to be a centralized operation financially, but the plantations would have "their own churches and schools" with each functioning as "an independent and self-sustaining community."[33]

In testimony before the Senate Committee on Education and Labor in September 1883, Calhoun explained in detail his assessment of the condition and needs of southern agriculture, alluding frequently and at length to his experience as a cotton planter in Chicot County. In a vein worthy of "New South" spokesman Henry Grady, he assured the committee that southern antagonism toward the North had ceased to exist, that the South was firmly committed to "the national idea," and that racial harmony prevailed throughout the region. The federal government, he insisted, should cease any agitation of "the so-called negro question" and should turn its attention to more constructive efforts in solving the South's problems.[34]

According to Calhoun, the most urgent need of the region was capital. The lack of capital not only forced many planters in the lower Mississippi Valley to become "nothing but agents of the [cotton] factors," but also bred a host of other evils, including exorbitant interest rates, burdensome debts, frequent foreclosures, and high prices charged by merchants. Calhoun urged federal action in four areas as a means of alleviating these conditions: abolition of the high tariff; establishment of postal savings banks; levee construction along the Mississippi River; and reduction of both direct and indirect taxes. But nothing, he maintained, would promote Southern agriculture more than an infusion of northern investment, which was unlikely so long as the federal government agitated the race question, because northern capitalists refused to risk money in "a country that might be involved in riots and disturbances."[35]

As Calhoun clearly indicated to the Senate committee, blacks occupied a central place in his vision of a prosperous agricultural South. Arriving in Chicot County just as Reconstruction entered its most turbulent phase there, he witnessed the "Chicot Massacre" of 1871, which he described to the Committee and cited as an example of the political agitation of the race issue that deterred northern capitalists from investing in plantation agriculture. Calhoun described race relations in Chicot County as entirely harmonious in 1883, even though blacks held virtually all county offices. In fact, he characterized the county's probate judge, a Negro and a tenant of the Calhoun Land Company, as "a very capable man, an excellent and good man, and a very just one." This judge had appointed Calhoun to transact "important business" for the county in New York.[36]

Based upon his experience, Calhoun was optimistic about the future of blacks in agriculture. He boasted that a few exceptional Negroes in Chicot County already had "their own factors in New Orleans, shipped their own goods, and received their own accounts of sales." He explained how those who accumulated "a surplus" would advance money "to another negro and take a mortgage on his crop." Those who possessed such a surplus, as he made clear, were rarely landowners, but rather were tenants themselves. According to Calhoun, it was to the advantage of white planters to encourage black tenants to forsake indolence, idleness, and extravagance in order to secure "the necessaries of life," including the ownership of a home, by which he meant a house and small plot of land. The assumption was that black homeowners would remain tenants on large plantations such as Sunnyside.[37]

In his discussion of blacks, Calhoun devoted special attention to education, pointing out that five public schools for Negroes existed on Calhoun Land Company property. Education, he maintained, was essential if blacks were to fulfill the duties of citizenship, advance economically, and be protected from those who had preyed on their ignorance in the past. While Calhoun believed that it would "take probably generations" for blacks, "as a class," to attain the standards of whites, he clearly surprised the Senate Committee by his forceful expression of "high hopes" for the capacity of Negroes to develop into useful, productive citizens, particularly as "responsible small farmer[s]" in the Mississippi Valley, provided they had access to education.[38]

T. Thomas Fortune, a militant black editor in New York, was so favorably impressed by Calhoun's views that he appended Calhoun's testimony before the Senate Committee to his *Black and White: Land, Labor, and Politics*, published in 1884. Understandably, perhaps, in view of the prevailing racial climate following Reconstruction, Fortune considered

Calhoun's testimony as exhibiting a spirit of fairness to blacks that was exceptional among white southerners and all the more so coming from the grandson of John C. Calhoun.[39]

But the younger Calhoun never suggested that altruism alone inspired his vision for the future of southern agriculture and the place of blacks in it. Rather, his was a pragmatic approach based upon his own experience and aspirations as a planter. "The development of my business," he told the Senate Committee, "is necessarily based upon the development of the negro and the cultivation of my lands." An indisputable fact, he argued, was that blacks would not "be able to compete with the Caucasian" for several generations. In the meantime, while they remained tied to the land as tenants, they not only should receive the rudiments of education, but should also be encouraged to follow the example and counsel of white planters in all economic matters. The Calhoun Land Company, he believed, provided them with an ideal apprenticeship for ultimately realizing their potential.[40]

On Sunnyside and their other plantations, the Calhoun Land and Florence Planting companies preferred to use the "tenant system," in which laborers provided their own teams and farm implements, but frequently were forced to adopt the "half-share system." John C. Calhoun explained the conditions that precluded their use of the tenant system and how his companies assisted blacks to rise from the half-share arrangement to tenant status:

> We wish to make small farmers of our laborers, and bring them up as nearly as possible to the standard of the small white farmers. But this can only be done gradually, because the larger portion of the negroes are without any personal property. We could not afford to sell the mules, implements, & c., where a laborer has nothing. Therefore the first year we contract to work with him on the half-share system, and require him to plant a portion of the land he cultivates in corn, hay, potatoes, & c. For this portion we charge him a reasonable rent, to be paid out of his part of the cotton raised on the remainder. In this way all of the supplies raised belong to him, and at the end of the first year he will, if industrious, find himself possessed of enough supplies to support and feed a mule. We then sell him a mule and implements, preserving, of course, liens until paid. At the end of the second year, if he should be unfortunate, and not quite pay out, we carry the balance over to the next year, and in this way we gradually make a tenant of him. We encourage him in every way in our power to be economical, industrious, and prudent, to surround his home with comforts, to

plant an orchard and garden, and to raise his own meat, and to keep his own cows, for which he has free pasturage. Our object is to attach him as much as possible to his home. Under whatever system we work, we require the laborer to plant a part of his land in food crops and the balance in cotton with which to pay his rent and give him ready money. We consider this system as best calculated to advance him. Recognizing him as a citizen, we think we should do all in our power to fit him for the duties of citizenship. We think there is no better method for doing this than by interesting him in the production of the soil, surrounding him with home comforts, and imposing upon him the responsibilities of his business.

Laborers purchased their "home comforts" and other supplies from stores operated by the Calhoun companies. Profit from the "mercantile portion" of his land enterprise constituted, as John C. Calhoun admitted, a significant part of the return on the investment of his company.[41]

At Sunnyside and their other plantations, the Calhouns employed all types of labor arrangements common to the area: wage, sharecropping, and rental. Their wage labor included a large contingent of "unskilled field hands," who earned from ten to twenty dollars a month in addition to lodging, board, and a garden spot provided by the company. Share-cropping took several forms, but the most common was that in which the Calhoun Land Company furnished the land, tools, house, garden area, pasturage—"every expense for making the crop and preparing it for market"—and the sharecropper performed all labor; then, the company and sharecropper divided equally the proceeds of the harvest. Another form of sharecropping employed by the Calhouns involved laborers who owned their own teams, implements, and other essentials for making a crop. In such cases the tenant received two-thirds or three-fourths of the crop depending on "the quality and location of the land." Under the rental system the Calhouns rented or leased land for eight or ten dollars an acre to those who possessed the prerequisites for cultivating it.[42]

It soon became evident, however, that the Calhoun enterprises were not fulfilling the expectations of their investors. Short cotton crops, coupled with the damage done by a devastating flood in 1882 and the failure of a single railroad to penetrate the area of their plantations, cast serious doubt on the success of what John C. Calhoun referred to as an agricultural "experiment."[43] Some residents of Chicot County who had originally been enthusiastic about the Calhouns' venture became disillusioned. One claimed that their land companies were "not interested in the commonweal." They would "not do anything for the improvement of the county" and prevented "good citizens from doing so." In his view Arkansas should enact laws to prevent the existence of such companies anywhere within its boundaries because they were an "evil unmitigated."[44]

By the mid-1880s the Calhouns were deeply in debt to eastern capitalists who had purchased the first mortgage bonds issued by the Calhoun Land and Florence Planting companies. Transferring his land holdings to the Florence Planting Company in 1884, John C. Calhoun, his wife, and her aunt, Julia Johnson, moved to New York City where within a few years he became a prominent figure on Wall Street closely identified with the development and financing of railroads in the South.[45] In 1885 the Florence Planting Company, under the control of Patrick Calhoun, defaulted on $133,305 worth of bonds. By court order in May 1885, Calhoun transferred most of the lands in Chicot County owned by the Florence Planting Company to the trustee of the bond holders. Deeds were delivered to the new owners the following January.[46] In December 1885, Patrick Calhoun, through the newly formed Chicot Planting Company, issued another series of first mortgage bonds amounting to $110,000 through the Corbin Bank. "Harwood," an island plantation of over 3,000 acres, served as collateral for these bonds.[47] Within four years Patrick Calhoun had divested himself of his Arkansas lands and had joined his brother in devoting his energies primarily to the reorganization and manipulation of southern railroads.[48] Both emerged from their Arkansas land ventures financially unscathed; in fact, their foray into land and planting companies may have enhanced their wealth.

Austin Corbin of New York, who had been a financial backer of the Calhouns from the beginning, ultimately came into possession of their Arkansas lands, including Sunnyside Plantation. In October 1886 Corbin assumed control of their network of plantations in Chicot County under a new company called the Sunnyside Company, which two years later sold bonds in the amount of $300,000 to pay off its "floating debt" and to make "additional improvements" on its plantations. Described as a man of "unusual vision and initiative," Corbin was president of the Corbin Bank in New York, but was known principally as a railroad organizer and financier. Closely associated with both the Long Island and the Philadelphia and Reading railroads, he had also initiated several land development projects in New York and a huge game preserve in New Hampshire prior to embarking upon his Sunnyside venture in Arkansas. His efforts to bar Jews from his Manhattan Beach resort and his role as secretary of the American Society for the Suppression of Jews in 1879 attracted considerable publicity.[49]

During the first few years after acquiring his Sunnyside properties, Corbin encountered a serious labor shortage, allegedly caused by the refusal of black tenants, who had long cultivated the land under local landlords, to work for a "foreign" company whose daily operation was in the hands of local managers or "overseers." As a result, much of the Sunnyside Company's tillable land remained "idle for a year or two" after Corbin assumed control of it.[50] In 1894, a year after Arkansas limited the convict-lease system, Corbin made a sharecropping contract with the state. According to this arrangement, the Sunnyside Company provided land, seed, teams, and implements, while the state furnished convicts for labor and "all their expenses." By this contract, each party, the company and the state, received half of the proceeds of the harvest. The number of convicts employed on the company's land varied from season to season, but in 1894 the average number was 251, and the harvest that year amounted to over a million pounds of lint cotton. A similar arrangement between the state and company prevailed in the following year, by which time a state "convict camp" had been established on or near Sunnyside Plantation.[51]

By 1895 Corbin had instituted substantial improvements at Sunnyside Plantation. His steamboat, the *Austin Corbin*, was anchored

Section hands working track of the Sunnyside Railroad, October 1893.
(Courtesy Mississippi Department of Archives and History.)

in Lake Chicot, and a telephone line linked the plantation with Greenville, Mississippi. The short Sunnyside Railroad had an engine and six box cars used primarily to transport cotton to and from his gin. One of the reasons for Corbin's popularity among the planters in the area was his interest in extending this railroad westward to Portland and Hamburg in adjoining Ashley County. Local merchants and planters in Ashley County hastily organized for the purpose of securing deeds to the thirty thousand acres of land in the county that Corbin required before construction of the railroad could begin. In March 1896 the Sunnyside and Western Railroad was incorporated; Corbin, who owned a controlling interest, was president, and the directors included his son-in-law, George S. Edgell, and several locally prominent planters. Construction was scheduled to begin on September 1, 1895. Local residents were all the more enthusiastic about this railroad because it promised to provide an east-west route that would intersect with the north-south Missouri-Pacific line.[52]

In the meantime, Corbin decided to settle Italian immigrants on Sunnyside Plantation. According to local lore, he was able to implement this colonization project because his eldest daughter had married an Italian count who used his influence to persuade "several hundred Italian families to emigrate to Arkansas." Corbin's daughter had not married an

Sunnyside Landing at daybreak, October 1893.
(Courtesy Mississippi Department of Archives and History.)

The *Austin Corbin,* Corbin's sternwheel steamboat, in Lake Chicot.
(Courtesy Mississippi Department of Archives and History.)

Italian nobleman but a French artist, René C. Champollian, who had committed suicide in 1886.[53] The actual circumstances under which Corbin succeeded in arranging for families from central and northern Italy to emigrate to Chicot County were more complicated and involved negotiating with an Italian immigrant agency in New York and with Italian diplomats.[54] He not only secured their assistance but also persuaded Don Emanuele Ruspoli, the Mayor of Rome, to become his Italian partner in this venture. On January 11, 1895, the *New York Times* announced that Corbin had arranged with the Immigration Bureau to settle Italian farmers at Sunnyside. It was, according to the *Times,* "the first of several such schemes to colonize the idle lands of the South."[55]

In 1894 Corbin subdivided 3,125 acres of Sunnyside Company lands into 250 twelve-and-a-half acre plots with houses described as "very simple but solid." The price of each plot of land with a house was $2,000 payable over 21 years at an annual interest rate of 5 percent on the unpaid balance. Prospective Italian settlers at Sunnyside agreed to these terms in a written contract that also stipulated that the Sunnyside Company would purchase, "if asked," all cotton raised on each twelve-and-a-half-acre plot "at current price . . . less the freight and expenses" which "were

Steamer, with barges, taking cottonseed at Sunnyside landing on the Mississippi River. *(Courtesy Mississippi Department of Archives and History.)*

not to exceed $1 per bale." Each contract also provided for the creation of an arbitration commission to settle any dispute that might arise between the Italian purchasers and the Sunnyside Company concerning contractual obligations. This commission was to consist of three arbitrators, one chosen by the Sunnyside Company, one by the Italian landowner, and a third "by the two arbitrators so named."[56]

When the first boatload of Italians destined for Sunnyside arrived at New Orleans on November 29, 1895, they proudly displayed their contracts, which, as one observer noted, were at best "quasi certificate[s] to the possession of American land." Since most of them did not have a "sou in the world," some observers in New Orleans questioned the legality of Corbin's scheme. The New Orleans *Daily Picayune* editorialized about the need for enforcing laws against contract labor, but was unsure whether Corbin's arrangements evaded these laws.[57]

Corbin's importation of Italians coincided with a campaign by railroads and chambers of commerce in Arkansas to attract immigrants into the state. But it was abundantly evident in their promotional liter-ature that Arkansans preferred immigrants of the "right kind," specifically those who were "prosperous, frugal, thrifty and sturdy." Opposition to the "new immigrants" from southern and southeastern Europe became especially pronounced in the wake of the lynching of eleven Italians in New Orleans in 1891. Because the Arkansas press viewed southern Italians in particular as undesirable, its initial reaction to the Italian colonists at Sunnyside ranged from skepticism to overt hostility. Some editors expressed concern that the Italians were imported to assist in the construction of Sunnyside and Western Railroad. Others betrayed their xenophobia by references to "Corbin's Dagoes." An editor in Monticello expressed relief that Chicot rather than his own county was the destination of this "lot of Dagoes." But hostility toward the Italians at Sunnyside appears to have dissipated when it became known that they were from central and northern Italy and were to engage in agriculture. In fact, a few other large planters followed Corbin's example, most notably John M. Gracie, the "largest individual planter in Arkansas," who imported Italians to work on his plantations in Jefferson County.[58]

The death of Austin Corbin a little over six months after the arrival of the first Italian colonists at Sunnyside prompted effusive eulogies by residents of southeast Arkansas who saw him as a benefactor of their region and as "a man of large means and enlightened and progressive views." But their primary interest in Corbin's "plans and investments" did not concern the colonization of Italians so much as it did his interest in building a railroad. Assurances from Corbin's son-in-law, Edgell, that the construction of the Sunnyside and Western Railroad would proceed on schedule temporarily dispelled their apprehension about the future of the project. Even more promising was the prospect, raised by Edgell, that the proposed railroad would extend westward to Warren, Arkansas, a total distance of seventy-five miles.[59] A visit to Sunnyside in the summer of 1896 by Ruspoli was interpreted as an encouraging sign because the Italian nobleman was viewed as one whose "vision" was similar to Corbin's. Later in the same year Edgell's representatives at Sunnyside entertained a large delegation of prominent planters and businessmen from south-eastern Arkansas at a banquet at the plantation headquarters. At Lake Chicot the group boarded the *Austin Corbin* and sailed around the lake where they were met by a rail car and transported to Sunnyside Landing. There the guests were entertained at "Corbin House." Promises and assur-ances given at this gathering encouraged local residents to believe that the Sunnyside and Western Railroad would soon be a reality.[60]

The enthusiasm inspired by what appeared to be the commitment of Corbin's heirs to complete the railroad project soon evaporated as delays in beginning construction followed one after the other. The refusal of the Arkansas legislature in 1897 to enact a measure conveying to the Sunnyside and Western one thousand acres of forfeited state lands for each mile of rail construction ended all hopes that the railroad would ever be built. Not until 1902 did Ashley and Chicot counties secure an east-west railroad, one financed principally by two "Colorado capitalists" and known as the Mississippi River, Hamburg and Western Railway.[61]

Although a second group of Italians had arrived at Sunnyside in January 1897, all was not well on the plantation. In fact, complaints about the climate, water supply, and malaria and other diseases common to the area, coupled with the "great dissatisfaction" regarding certain features of their contracts, prompted a sizable contingent of the Italians at Sunnyside to migrate in 1898 to northwest Arkansas on the border of the Indian Territory, where they founded Tontitown. Others settled in Missouri. The conditions at Sunnyside, along with the death of Corbin,

A bridge across a bayou near the landing house, showing cotton that has been sampled.   (*Courtesy Mississippi Department of Archives and History.*)

contributed to the decision of the Corbin heirs to lease their plantations. In December 1898, the Sunnyside Company leased three of its plantations, Luna, Patria, and Latrobe, for five years to Gayden Drew, a local planter, for an annual rent of $3,300. The other four, Sunnyside, Fawnwood, Hebron, and Hyner were leased to Hamilton R. Hawkins and Orlando B. Crittenden, merchants and cotton factors, and to Leroy Percy, a planter and politician, all of Greenville, Mississippi, who agreed to pay the Sunnyside Company an annual rental "equal to one half of the [net] profits of a year's planting operations."[62]

The publicity about the "experiment" in Chicot County generated by Percy and others proclaimed that the Italian, because of his superiority to the Negro as a cotton grower, offered the solution to the South's labor problem. Despite such glowing reports about the "Italian experiment," conditions on the plantations proved to be considerably less idyllic than the publicity suggested. The number of Italian families on the plantations not only dramatically declined, but the Sunnyside enterprise also experienced other difficulties, including floods, an infestation of the boll weevil, and well-publicized accusations of peonage. The failure of the Sunnyside Company either to realize the income anticipated by its investors in 1888 or to sell the plantations for a profit forced it to issue a second series of bonds for $400,000 in 1910 in order to retire its earlier bonds with interest.[63] In the next decade blacks replaced Italians as the plantations' principal labor force, and sharecropping became the prevailing labor arrangement.

In May 1920, as the price of cotton soared and the value of delta land dramatically increased after World War I, the Sunnyside Company sold its Chicot County lands for almost one million dollars to W. H. and J. C. Baird. The Bairds, who had executed notes for five hundred thousand dollars and assumed the payment of all outstanding Sunnyside Company bonds, defaulted less than a year later when cotton prices dropped precipitously. In 1924, the Kansas City Life Insurance Company, as representative of Sunnyside Company bond holders, sued the Sunnyside Company and the Bairds. When the court ordered that Sunnyside lands be sold at public auction to the highest bidder, Kansas City Life Insurance Company purchased them for $75,000. In 1926 Kansas City Life sold its plantations to the Jewell Realty Company of Kansas City, a transaction that resulted in foreclosure, leaving the Sunnyside lands in the possession of the insurance company. Although local school, drainage, and levee districts had claimed portions of the Sunnyside property, it still contained 8,463.4 acres in 1930.[64] Rumors that

Henry Ford had dispatched a representative to inspect the property with a view toward purchasing it for raising cotton to be used for automobile upholstery appeared to enhance the prospect for Kansas City Life to sell its Chicot County land. But Ford never made an offer, and the insurance company remained the owner of Sunnyside.[65]

In the late 1930s when employees of the Federal Writers' Project went to Chicot County to collect data, they found that there was little evidence that Sunnyside had once been a thriving agricultural enterprise. Austin Corbin's Sunnyside Railroad and all of its rolling stock had "long since been removed"; the steamboat on Lake Chicot had been scrapped and sold; fires had destroyed the cotton gin as well as the Sunnyside Company store and other buildings. Only two Italian families remained a part of the plantation. In a little less than a century after Abner Johnson sold Sunnyside to Elisha Worthington in 1840, the plantation that once had been the showpiece of one of Arkansas's wealthiest delta counties had become home for "clients of the New Deal's Resettlement Administration."[66]

Actually, Kansas City Life Insurance Company leased Sunnyside to the Arkansas Rural Rehabilitation Corporation in 1935 for a period of three years with an option to purchase the property for $150,000. The lease provided that each of the 123 ARRC tenants was to plant a minimum of eight acres of cotton and that Kansas City Life Insurance Company was to receive one-fourth of the sale price of all cotton and cotton seed "after deducting the usual cost of ginning." Since the ARRC did not exercise its purchase option, control of Sunnyside reverted to the insurance company in 1938. Between 1941 and 1945 the company sold, in tracts of various sizes, most of the acreage of the plantations that had been combined by the Calhoun brothers in the 1880s and that had been known ever since by the name of the largest single component, Sunnyside Plantation.[67]

Carved out of a swampy, heavily wooded wilderness early in the nineteenth century, Sunnyside Plantation was a product of the westward movement of cotton agriculture. In the century following its formation, the plantation experienced, sometimes in dramatic form, the developments that shaped the southern economy—fluctuations in cotton prices occasioned by changes in the world demand for cotton, destruction and turmoil produced by the Civil War and Reconstruction, the shift from slave to free labor, the invasion of the boll weevil, and federal policies relating to flood control and agricultural production. A succession of well-known individuals, some bearing famous names, figured, either directly or indirectly, in the checkered history of Sunnyside. Ownership of the complex, largely self-sufficient enterprise created by Elisha

Worthington ultimately passed to an absentee corporation. The name Sunnyside thereafter applied to a combination of large plantations cultivated, as before the Civil War, by blacks except for an experiment with Italian labor that not only failed but also led to charges of peonage and aroused the enmity of the Italian government. In view of the plantation's past, the reference by a Federal Writers' Project employee to "shadows over Sunnyside" was more appropriate than he probably realized.[68]

Labor Relations and the Evolving Plantation: The Case of Sunnyside

Jeannie M. Whayne

The transformation of the Sunnyside Plantation in the late nine-teenth and early twentieth centuries took place within the context of the evolving southern plantation system following the Civil War. It was a metamorphosis that ultimately witnessed the advent of agribusiness and a reliance on capital-intensive rather than labor-intensive methods of cul-tivation. The implications of this great change necessarily involve both economic and social questions of significant relevance for those inter-ested in the history of the South. The choices planters made concerning how to finance the crop and modernize their operations, even as they faced monumental obstacles in the form of declining cotton prices, nat-ural disasters, and labor problems, are matters that historians have sub-jected to close scrutiny. The decisions that freedmen made, even though those decisions were circumscribed by limited access to landownership, the political resurgence of the planter class, and tangible liabilities in the form of illiteracy and poverty, are also subjects that historians have stud-ied and analyzed. Although the debate is by no means exhausted, it has produced a substantial body of historical literature that addresses certain essential questions.

Was the South taken advantage of by a vengeful North during the hated Reconstruction period, or did the North squander its oppor-tunity to truly "reconstruct" the South and then relinquish its moral obligation to the freedmen? Were the planters displaced, or did their landholdings help them weather the storm that was Reconstruction?

Were the freedmen so ill prepared for freedom that they could not hope to make the most of it, or did they have sufficient opportunity to fully exploit their new status? Did paternalism survive the end of slavery, or did it wither on the vine as planters embraced capitalist agriculture? Indeed, was there a free market in labor or did certain institutional constraints delay the emergence of capitalist agriculture? And, finally, were disfranchisement and legalized segregation manifestations of a new and virulent form of southern racism, or were they merely strategies used by white elites to divide the black and white lower classes and thus solidify planter predominance?

Historians have addressed these questions in a variety of ways and have come up with a multiplicity of answers. At almost every juncture, moreover, Willard Gatewood's study of the Sunnyside Plantation's transformation provides a unique window on the evolution of the southern plantation in the post–Civil War period. Elisha Worthington's experience during Reconstruction in Chicot County sheds light on the extent to which a particular planter was affected by the war and its aftermath; moreover, the ability of local planters to regain control after Reconstruction officially ended in Arkansas demonstrates the limited commitment of the national Republicans and provides at least one answer to the question of whether planters were displaced. The adoption of sharecropping and tenancy at Sunnyside and the evolution of Arkansas lien laws that worked to the benefit of landlords are indications of the constraints placed upon freedmen and makes readily apparent why planters soon faced a severe labor shortage as freedmen left the area in search of better opportunities elsewhere.[1] The emergence of a form of capitalist agriculture, as planters tried "creative financing" to keep their operations afloat, helps explain the rise of absentee ownership and thus the demise of paternalism. The experiment with Italian immigrant labor throws light on the meaning of disfranchisement and segregation, for the Italians were, on the one hand, noncitizens who could not vote and, on the other, a "third race" that experienced de facto segregation from whites and scant solidarity with blacks.

Some of these questions hinge on the political implications of the transformation that took place after the Civil War. But even the studies that were primarily concerned with politics rather than with economics had an economic component. The Gatewood overview of the Sunnyside Plantation provides an opportunity to examine the economic component of the history of Sunnyside. Significantly, historians who focus on the economic ramifications of the plantation's transformation either implicitly or explicitly deal with one crucial factor: the emergence of capitalist agriculture in the South. Some historians have concluded that the sharecropping system of labor was not free labor and that planters behaved more like the landed elite of the antebellum period. For them, capitalism did not emerge. And while there is agreement among other historians that capitalism flowered in the South after the Civil War, there is considerable disagreement over how that growth affected both the planter class and those who worked the plantations. These are by their very nature "economic" questions, but the historical literature on Reconstruction truly begins with the more traditional political studies, studies, once again, that necessarily involved economic questions.[2]

For generations the prevailing view of Reconstruction depicted the South prostrate before a vengeful North. Historian William A. Dunning was so much a part of initiating this dominant view that his name came to be identified with it, and thus historians speak of the Dunning interpretation or the Dunning school.[3] Even in its heyday, however, certain historians began to raise questions about the capitalist nature of the South after the Civil War. Prominent among those historians was Charles Beard who argued that the Civil War itself was "the second American revolution." For Beard, the Civil War was a contest between the industrial capitalists of the North and the planter aristocrats of the South.[4] Yet that point of view failed to dislodge the prevailing Dunning interpretation that focused on the South as victim rather than subjecting the new economic system to substantive analysis. Thus the Dunning school reigned supreme until the 1960s, when historians, writing during the Civil Rights movement, began to ask whether the Civil War meant anything at all given the limited progress blacks had made in the South. But even before the Civil Rights movement transformed the terms of the debate, C. Vann Woodward delivered a devastating blow to the traditional view with two books, both published in 1951. In *Reunion and Reaction: The Compromise of 1877 and the End of Reconstruction* and in *Origins of the New South: 1877–1913*, Woodward truly redefined the terms of the debate and inspired a whole new generation of historians.[5] He argued that the compromise following the controversial election of 1876 preserved "the pragmatic and economic parts" of Reconstruction at the expense of "the idealistic and humanitarian parts."[6] Both North and South were facing serious labor problems and, in any case, the Republican party had changed over the years so that it was more sympathetic to business interests than to Radical Republican policy.[7] Even the Fourteenth Amendment itself began to be interpreted in ways that served the needs of corporations rather than freedmen. The publication of these two books prompted a thorough re-examination of the Republicans and the reasons for the failure of Reconstruction.[8]

But Woodward's *Origins of the New South* spawned yet another line of questioning that has preoccupied historians over the last four decades. He subjected the so-called Southern Redeemers to a searching analysis, found that they courted northern capital, and concluded that the Redeemers "were of middle class, industrial, capitalist outlook, with little but a nominal connection with the old planter regime."[9] Their courtship of northern capital, especially as it related to their efforts to refashion the railroad system in the South, brought disaster to many of them during the economic crisis beginning in 1873. Not only did many of the Redeemers have middle-class origins, but they successfully integrated themselves into the mainstream of the southern political elite. Woodward did not argue that the planters were excluded or thoroughly overturned by these new (southern) men from the middle class, but rather that planters had to share center stage with them and that this fundamentally affected the way that the South evolved in the years following Reconstruction. This line of thinking encouraged historians to investigate just how devastated planters were.

James L. Roark in *Masters Without Slaves: Southern Planters in the Civil War and Reconstruction* argued that "those planters who were resilient, adaptable, and skillful enough to diversify their investments and to find sound commercial and industrial opportunities were more likely to retain their plantations."[10] Jonathan Wiener in *Social Origins of the New South: Alabama, 1860–1885* found that a significant number of Alabama black belt planters survived the Civil War and that the key to their survival was their ability to hold on to the land they owned.[11]

Worthington and his compatriots in Chicot County exemplify the ability of planters to successfully adapt to the profound transformation confronting them. Even in a state of declining health, Worthington possessed enough business acumen to salvage his enterprise and to avoid economic disaster. In order to pay debts he had incurred during the 1850s, he sold Sunnyside in 1867 to an old Kentucky acquaintance, Robert P. Pepper. But he retained possession of his other three plantations and thus preserved much of his economic clout. Along with several other planters in Chicot County, Worthington somehow used the bankruptcy law of 1867 to circumvent economic ruin, and his enterprise, though diminished, survived until his death in 1873.[12]

Sunnyside itself, purchased by Lyne Stalling from Pepper, experienced a revival in the mid-1870s under the management of the William Starling Company. It was an attractive enough investment to encourage John C. Calhoun, the grandson of the famous South Carolinian by the same name, to purchase it in 1881. In conjunction with his brother, he incorporated in Connecticut and operated Sunnyside and his other plantations in Chicot County under a succession of land and planting companies. He argued for diversification and the introduction of business methods. Even more than Worthington, perhaps, Calhoun sounds like the kind of planter historian James Roark had in mind when he wrote about "resilient, adaptable, and skillful" planters who adjusted to new realities in the region. Although the Calhouns found themselves so deeply in debt by the mid-1880s that selling out was their best option, they emerged from their Chicot County enterprise economically unscathed.

What of the freedmen of Chicot County? The historical literature on the fate of freedmen in the South in the post–Civil War era is extensive, and the Chicot County example is illuminating. Historians began to re-examine the experience of the freedmen in the early 1960s, and no historian is perhaps more important to this historiographical trend than Willie Lee Rose. In *Rehearsal for Reconstruction: The Port Royal Experiment*, she argued that in failing to confiscate and redistribute plantations, Republicans failed to give freedmen the foothold they needed to survive in the new world into which they were being thrust.[13] Rose was essentially writing from the black perspective, a perspective taken by only a few previous historians, most notably, Alrutheus Ambush Taylor, Carter G. Woodson, and W. E. B. DuBois. Rose's work was followed by that of Louis E. Gerteis in *From Contraband to Freedmen: Federal Policy Toward Southern Blacks, 1861–1865*.[14] Gerteis pointed out that the Civil War did not make possible the kind of sweeping social change that blacks needed in order to avoid becoming slaves by another name. Northern policy during the Civil War was limited by the desire to avoid social revolution. This policy foreshadowed that implemented in the postwar period.

When Worthington returned from Texas at the end of the Civil War, he found that his plantations were listed as "abandoned lands" by the Freedmen's Bureau. Worthington had no sooner returned with his former slaves to Arkansas than the Freedmen's Bureau imposed the wage labor system. Soon the Freedmen's Bureau was disbanded, and Worthington turned to the new share system emerging in the South. So even though he sold Sunnyside in order to salvage his enterprise, he adapted to the realities of the new labor system and survived.

Yet another intriguing fact emerges from the Gatewood study of Sunnyside that relates to the use of freedmen as sharecroppers. James Mason, Worthington's son by a slave woman, actually ran at least one of his father's plantations, as well as lands that he acquired on his own, and he utilized freedmen as sharecroppers. The complicated relationship between Mason and his workers remains a mystery given the dearth of information about precisely how Mason related to his sharecroppers, but the fact that he employed freedmen in this manner suggests that complex factors were at work in the evolving plantation system in the years following emancipation.

Whatever insight the Mason case might provide, it is likely that his sharecroppers, his father's former slaves, and other freedmen in the area learned the lessons Joel Williamson outlines in *Crucible of Race: Black-White Relations in the American South since Emancipation*. Although blacks secured political rights during Reconstruction, the franchise was not enough to provide them with sufficient leverage to counterbalance the economic clout that planters could muster.[15] With economic control came political control.[16] It took some time, however, for whites to regain political dominance on the local level in Chicot County. When Reconstruction ended in Arkansas in 1874, Chicot County blacks continued to hold local offices, and as late as 1883 the black probate judge was described by Calhoun as "a very capable man, an excellent and good man, and a very just one." Later in the decade, the positions of the black politicians in the county eroded, and by the time of the imposition of disfranchisement measures in the 1891 legislature, their influence had profoundly diminished. But the inability of freedmen to move beyond sharecropping in Chicot County was already an established fact by the time local blacks lost their power.[17]

By the 1880s planters in Chicot County were beginning to experience the difficulties of maintaining a so-called free labor system once blacks became more fully aware of how limited their opportunities were. Jay R. Mandle in *Roots of Black Poverty: The Southern Plantation Economy After the Civil War* examined the survival of the plantation, how planters asserted their control, and how very few options were open to freedmen.[18] It was work the plantations or nothing. But one option the freedmen continued to have was a certain amount of mobility. Institutional constraints hampered their movement, however. Harold Woodman in "Post Civil War Agriculture and the Law" demonstrated how lien laws evolved in such a way as to undermine the freedmen (and, by the way, to undermine merchants trying to operate in competition with planters).[19]

Finally, Pete Daniel in *The Shadow of Slavery: Peonage in the South, 1901–1969* showed how debt kept the blacks shackled to the plantation system.[20] Despite the operation of lien laws and emergence of debt peonage, blacks did find ways to move from plantation to plantation. Labor scarcity, therefore, became a major concern of local planters.

Some historians have argued that mobility worked to the advantage of the freedmen. Robert Higgs, for example, in *Competition and Coercion: Blacks in the American Economy, 1865–1914*, pointed to the significant gains made by blacks in the postbellum period, but any gains would have seemed spectacular given where the freedmen started. Higgs wanted to believe that the free market economy worked, and he argued that the ability of blacks to migrate to better opportunities was an important element in their economic survival.[21] Stephen J. DeCanio in *Agriculture in the Postbellum South: The Economics of Production and Supply* wrote with the same motives that inspired Higgs and came up with conclusions that are not dissimilar.[22] Rather than looking at mobility as a measure of the freedmen's ability to survive economically, however, DeCanio used econometric techniques to focus on productivity. He concluded that black farmers were not, in fact, exploited and that they received a fair return for their labor.[23] Gavin Wright, meanwhile, in *Old South, New South: Revolutions in the Southern Economy since the Civil War*, returned to the question of the labor market raised by Higgs and, like Higgs, concluded that it worked to the advantage of blacks. An economist by training and inclination, Wright dismissed the importance of the institutional constraints outlined by both Woodman and Daniel.[24]

The studies of Higgs, DeCanio, and Wright rest on the assumption that the Civil War truly meant something and that "something" was, at least in part, the growth and development of a capitalist ethic in the postbellum South. Most historians of the South, whether or not they accept the conclusions reached by Higgs, DeCanio, and Wright, would agree that some kind of capitalism did emerge. Jonathan Wiener's *The Social Origins of the New South* most forcefully took issue with this contention. He argued that the South took the "Prussian road" to economic development. According to Wiener the development of the South diverged significantly from the capitalist path and continued its precapitalist origins into the postwar period. Dwight Billings, meanwhile, in *Planters and the Making of a 'New South': Class, Politics, and Development in North Carolina, 1865–1900* believed that capitalism developed in the South but that it was a peculiar kind of capitalism. Because he found planters and planter capital playing a dominant role in post–Civil War North Carolina, he argued that industrial development was "peripheral" to the agricultural economy there.[25] Although he did not go as far as Wiener, Billings did insist that the South took a qualitatively different path to development.

But the Wiener and Billings arguments met considerable opposition. Eugene Genovese and Elizabeth Fox-Genovese, in *The Fruits of Merchant Capital*, for example, insisted that capitalism did take hold and that the paternalistic system in existence before the Civil War gave way as planters "eventually agreed upon bourgeois principles of freedom of labor."[26] Genovese and Fox-Genovese did not try to argue that the system worked to the advantage of blacks, however. But they raised another question that has been a source of puzzlement since the end of the Civil War: the transformation of the paternalistic impulse. Eugene Genovese is particularly qualified to address this question because he articulated the most persuasive and influential view of antebellum paternalism in *Roll, Jordan, Roll: The World the Slaves Made.*[27] According to *The Fruits of Merchant Capital*, "[E]ven the most admirable and genuinely paternalistic of the old landed classes generally surrendered the best of their traditions, most notably the organic view of society and the idea that men were responsible for each other, while they retained the worst of their traditions, most notably, their ever deepening arrogance and contempt for the laboring classes and darker races."[28] But in the transition period immediately following the Civil War, the remnants of paternalism often helped planters keep their plantations running. Jay R. Mandle in *Roots of Black Poverty* placed greater emphasis upon the survival of paternalism, but argued that it was a different kind of paternalism. For Mandle, "Economic calculation assumed a new importance" and the "frequent use of violence" constituted "evidence of ideological weakness . . ." in the postbellum period.[29] For Joel Williamson, meanwhile, "The paternalistic attitude of whites gave way to bitterness" but paternalism itself survived "among white liberals. . . ." Williamson even saw it emerging in the mill villages that arose in the South after the Civil War.[30]

Again, the case of Sunnyside provides a unique window on yet another historiographical debate. When Worthington employed his former slaves as sharecroppers on his remaining plantations, he probably continued his paternalistic interest in them. But by the time the Calhouns took over the enterprise, what paternalism remained was most certainly disintegrating. The evidence clearly suggests that the Chicot County freedmen resented "the companies" and perhaps their resentment reflects the fact that "the companies" had accepted "bourgeois principles of freedom of labor" and abandoned "the organic view of society and the idea that men were responsible for each other . . ."[31] Once Austin Corbin purchased Sunnyside, any semblance of the paternalistic system employed by the likes of Worthington had completely disappeared. Many blacks simply refused to work for the "foreign" company,[32] and Corbin resorted to the use of convicts to keep his enterprise running.

The Sunnyside example does not provide clear evidence of the virulent form of racism that emerged in the South in the late nineteenth century, but certainly the freedmen there must have experienced the detrimental effects of segregation and disfranchisement. Williamson's *Crucible of Race* focused on the evolution of racism in the postbellum South—an issue that is necessarily bound up with the emergence of segregation and disfranchisement after the demise of slavery. C. Vann Woodward's *The Strange Career of Jim Crow* remains one of the most significant works on the imposition of segregation.[33] Woodward argued that certain restraining forces (northern liberals, southern conservatives, and southern radicals) protected the position of the freedmen in the post–Civil War period. The turn to extreme racism and the imposition of segregation (and disfranchisement) was due not to a *conversion* of these restraining forces but rather "to a relaxation of" their opposition. The southern radicals, largely through the Populist movement, had seriously challenged the ruling Democratic party in the South and threatened to unite blacks and whites and displace the Democrats. The southern conservatives, in control of the Democratic party machinery, recognized the threat of the Populists and the failure of their own efforts to successfully incorporate the black electorate into the Democratic party through a strategy known as fusion. Such a strategy had been designed to lure blacks away from the Republican party with promises of minor positions within the Democratic machine. But this was always a difficult marriage, for the southern conservatives had used the race issue, at least in part, to regain control of the political process from the Republicans in the first place, and their courtship of blacks seemed incongruous to many white southerners. When the radical challenge emerged in 1890, the conservatives essentially panicked, abandoned blacks to their own devices, and utilized racist rhetoric to fight off the Populists.

Joel Williamson differed with Woodward's view of the imposition of segregation. He argued that blacks initiated separation immediately following the Civil War in an effort to distance themselves from the scenes of old injuries and injustices. They created their own communities and established their own churches apart from the white world. But blacks were not living in isolation. A constant interplay between the white and black worlds in the years following emancipation influenced the evolution of southern society and affected the ideas that blacks and whites had about each other. This interplay clearly influenced the reshaping of the white world, and, according to Williamson, whites were distinctly uncomfortable with the black influence. Certain whites believed blacks were retrogressing toward a "natural state of savagery and bestiality," and that this would have profound consequences for the white world. By the end of the nineteenth century, whites began to use racism to limit this imagined negative influence.[34]

A work which opens up the debate in a new and promising way is Loren Schweninger's *Black Property Owners in the South, 1790–1915*.[35] In his comprehensive and thoroughly researched study, Schweninger examined forty-one thousand black property owners in the South from the end of the eighteenth century to the beginning of the twentieth century. According to Steven Hahn, property ownership among blacks was "more widespread and substantial among nineteenth century blacks than previously assumed."[36] The post–Civil War black owners, however, faced certain obstacles, and racism played no small part in their problems. They found themselves on the periphery and dependent upon the sponsorship of powerful whites in order to survive.

William Cohen, meanwhile, in *At Freedom's Edge: Black Mobility and the Southern White Quest for Racial Control, 1861–1915*, recently re-examined the question of mobility and, according to Hahn, "sees a close relationship between white efforts to restrict black mobility and the onset of segregation and disfranchisement."[37] In arguing that the seeds of segregation and disfranchisement were sown at least two decades before the end of the nineteenth century, Cohen thus challenged Woodward's "periodization." He did argue, however, that white supremacist ideas (and thus racism) constituted a significant factor in the phenomenon.

J. Morgan Kousser in *The Shaping of Southern Politics, Suffrage Restriction and Establishing of the One-Party South, 1889–1910* focused on disfranchisement rather than segregation and found that the threat posed by the Populists beginning in the early 1890s played a major role in the disfranchising measures adopted by southern states.[38] Fearing the unification of the black and white masses, Democrats eliminated both from participation in the franchise. The Democratic party was aided by the realization on the part of northern Republicans that they no longer needed their southern counterparts to hold on to the reigns of power on the national level. Thus the national Republican party turned its back on the South. By the 1890s northerners in general were far more interested in national reunification and in imperialistic designs abroad supported by both national parties.

In the meantime, northerners were themselves facing hordes of eastern Europeans who posed a threat to their own hegemony, and thus identified with the southern whites who regarded blacks as a negative influence. The treatment of the Italian immigrants at Sunnyside during this crucial period was hardly likely to inspire great concern on the part of the federal government or northerners in general. But added to the fact that the Italians were a "third race" between blacks and whites was the fact that they were poor, Catholic, and spoke little or no English. While the latter two handicaps have received little historical analysis by historians of the South (given the dearth of either immigrants or Catholics in the South), the discussion of the condition of the poor in the South has been regenerated by the recent work of Barbara J. Fields.

In an essay appearing in *Region, Race and Reconstruction*, Barbara Jean Fields took the debate into waters charted by only a few historians. She argued that historians have looked too narrowly at racism in an attempt to explain the black condition.[39] She hardly denied that racism existed, but suggested that it is not an adequate analytical tool; instead, she insisted, the important issue is *class* not race. William Cohen in *At Freedom's Edge* disputed the notion that class was the significant factor and argued, inexplicably, that his work challenged the "prevailing interpretation" that "emphasizes class factors and treats the effort to segregate, disfranchise, and control blacks as part of a larger class conflict between planters and other moneyed interests, on the one hand, and poor farmers, both black and white, on the other."[40] Unfortunately, Cohen did not identify the "prevailing interpretation" he said he challenged.[41]

Certainly, given the nebulous racial category of the Italian immigrants, class may well be a better way of interpreting their experiences at Sunnyside. And if their poverty was not enough, their religion, their language, and their foreignness set them apart in a culture based upon a network of kinship relationships. In this context, their darker hue was just one more handicap. Even though the local whites were initially relieved to hear that the Italians coming to Sunnyside were from northern rather than from southern Italy (and thus more likely to be "the right kind" of immigrant), once they arrived on the scene, the white community refused to accept them.

As indicated by the recent work of Schweninger and Cohen, historians of the South continue to debate certain of the questions raised earlier in this essay. Perhaps the most controversial issue has to do with the nature of the capitalist ethic taking hold in the South after the Civil War and what that meant for those who labored on the plantations. The individuals who worked those plantations, whether they were native blacks and whites or whether they were Italian immigrants, found limited opportunities, and despite their best efforts, few were able to work themselves out of debt and into landownership. Indeed, the tendency toward sharecropping and tenancy accelerated in the years after the turn of the century as more land became concentrated in fewer hands. The Italians at Sunnyside arrived just as the trend was shifting decidedly away from small farm owners and toward large plantations. In this sense, the Sunnyside experiment was an attempt to challenge a powerful trend in southern agriculture, a trend that the Italian immigrants were ill equipped to transcend.

PART TWO

Peonage at Sunnyside and the Reaction of the Italian Government

Ernesto R. Milani

The history of Italian-American labor relations in the late nineteenth century is marked by a number of ironies. Both the fragmented Italian states and the American South had been forced by war and political upheaval into new unities. The Civil War and the Risorgimento promised a dawning of a new era for the humble fieldworkers so long oppressed under exploitative labor arrangements. But the day of jubilee clearly had not arrived by the turn of the twentieth century. While thousands of disillusioned peasants fled the Mediterranean to seek their fortunes in countries such as Brazil, Argentina, Canada, and the United States, American black laborers also began to leave the land. Less mobile than the European peasants, blacks in the South sought job opportunities in the cities, sawmills, and mines of the region of their birth.

Many southern landholders, particularly in the soil-rich Delta, grew alarmed at a mounting labor shortage. Moreover, they were frustrated by the seeming backwardness of the black sharecroppers whom less demoralized compatriots had left behind on the farms. The landholders' answer to the South's labor problem was to import Italian immigrants to work the region's cotton fields. The effort came close to reducing the Italians to the same form of debt servitude that kept so many blacks tied to the same land their fathers had worked as slaves. The Italian government's inattention to the plight of its own citizens was much like the North's loss of interest in the plight of blacks after Reconstruction. The story of the Sunnyside Plantation reveals not only the insensitivity of the Italian authorities, but also the indifference of the southern landlords to the needs and dreams of this new class of workers.

In 1894 Austin Corbin, a banker, land speculator, and organizer of the Long Island Railroad, approached his friend Alessandro Oldrini, chief agent of the Bureau of Information and protection for Italian Emigration at Ellis Island, about a scheme to populate the Sunnyside Plantation in the Arkansas Delta with Italian immigrants. He described the enterprise to Oldrini as a semi-Utopian experiment that would bring prosperity to the immigrants. In fact, Corbin needed to raise capital by selling a portion of his mortgaged plantation. His plan was to subdivide 3,200 acres of Sunnyside into twelve-and-one-half-acre lots and sell them at $2,000 each to approximately 250 Italian families. The Italians were supposed to cultivate cotton and from their profits make payments on their plots over a twenty-year period before taking title.[1]

The Italian ambassador to the United States, Francesco Saverio Fava, did not officially sanction Corbin's enterprise, but neither did he interfere with its implementation. He simply made it clear that the Italian

Railway station of Tavernelle, where the first would-be cotton growers of Recoaro left for Sunnyside. *(Private Collection of Ernesto R. Milani.)*

government considered it a private transaction between Corbin and the future colonists. The Italian embassy did, however, sponsor a visit to the plantation in October 1894 by Ambassador Fava's son and Don Emanuele Ruspoli, the mayor of Rome. When they visited in the fall, they found the climate temperate and the area mosquito free. They also noted that the plantation had such promising features as its own railroad and rolling stock, a steamboat, a store, gin, sawmill, grist mill, and several new farmhouses. Agricultural implements were ready and waiting for the newcomers. Ruspoli was so favorably impressed that he became Corbin's Italian partner and agreed to recruit what he hoped would number 250 Italian families.[2]

The *Chateau Yquem,* carrying 98 Italian families from Marche, Emilia, and Veneto, docked in New Orleans on November 29, 1895. There were 303 adults, 110 adolescents, and 127 children.[3] Oldrini, who was probably on Corbin's payroll, met the group and brought them to Sunnyside. He and the Italian workers soon realized that the immigrants were faced with a number of difficulties. First, the Sunnyside managers did not speak Italian and would not be able to understand the immigrants' concerns and needs. Furthermore, the immigrants found that the Sunnyside Company had less extra work available than had been promised. They would have to rely on company advances—half in cash and half in commissary paper—for their livelihoods until the first cotton crop came in. Some immigrants complained that there was no price differential for less productive land being sold to them. The Italians also discovered that the water supply was unsanitary, and as the hot weather began in the summer of 1896, so did the swamp fevers that would ravage the colony. The installation of filters was the only remedy offered by the company in spite of requests to provide artesian wells. In a report to the Italian embassy, Father Pietro Bandini, a priest from New York assigned to Sunnyside, confirmed the problems observed by Oldrini.[4]

In June 1896, as the immigrants were beginning their first crops, Austin Corbin was killed in a carriage accident in Newport, New Hampshire. The Italian Royal Ministry of Foreign Affairs viewed his death with apprehension and feared for the future of the Sunnyside experiment. However, after a favorable report by Ruspoli on his second visit to Sunnyside made later that summer, Ambassador Fava agreed to permit the embarkation of a second group of seventy-two families from Genoa on December 17, 1896. They arrived at Sunnyside on January 5, 1897, and learned that the earlier group of immigrants had experienced mixed results with their first cotton crops. Over one-third of the

New ground cotton. Cultivated by Italian settlers near Sunnyside, Arkansas.
(Private Collection of Ernesto R. Milani.)

families remained in debt to the company after expenses for their voyage, supplies, equipment, and furnishings, in addition to advances and interest, were deducted from their cotton receipts.[5]

By May 1897 dissatisfaction had mounted, and the Italians had formed a committee headed by Giovanni D'Elpidio to present their complaints to the company. The colonists charged that they had been misled about the climate, the water, the availability of extra work for cash, and the high prices and interest rates charged at the company store. The cost of the lots at $160 per acre surpassed by 50 to 60 percent the price of the most productive land in the area, and the primitive houses were overvalued at a charge of $150. They demanded artesian wells and a change in the contract from the sale of the 12.5-acre lots to their rental, a revision that they thought would lower their indebtedness.[6]

At this point Ambassador Fava took an extended leave of absence and delegated to his subordinate, Giulio Cesare Vinci, the difficult task of dealing with the situation. The Royal Ministry of Foreign Affairs in Rome had played down the reports about the adverse conditions at Sunnyside made by Father Bandini to Ambassador Fava. The ministry claimed the promises of extra work were against contract labor law and that Count Ruspoli of Rome had assured them that artesian wells were under construction. Ruspoli had been a friend of Austin Corbin's and had made favorable reports on the Sunnyside experiment. To protect his reputation and credibility, he stalled authorities in Rome by attributing the colony's problems to the lengthy liquidation of Corbin's estate and the consequent inability of his heirs to take any action before they were legally authorized to do so.[7]

Father Bandini strongly criticized the Italian embassy for refusing to order an on-site investigation of Sunnyside. An epidemic of yellow fever had left forty-four children and twenty-eight adults dead at the end of 1897. In spite of a large cotton crop the year before, a number of colonists voiced their desire to secure employment elsewhere, and the committee representing the immigrants renewed its complaints to the Italian embassy.[8] The acting ambassador realized after meeting with George S. Edgell, Corbin's son-in-law and manager of the estate, that an investigation was warranted. He was unsure, however, about the embassy's authority to take such action without instructions from Rome.[9]

Accounts of the conditions at Sunnyside appearing in the press prompted the Italian government in January 1898 to order Guido Rossati, an agronomist and director of the Royal Italian Oenological Bureau in New York, to investigate Sunnyside.[10] By the time he arrived the

Vaucluse cemetery where Italian immigrants to Sunnyside were buried.
(*Private Collection of Ernesto R. Milani.*)

following month, conditions had deteriorated. New practices, such as charging for services that previously had been furnished free, had further strained the relationship between the colonists and management. Rossati was not impressed by the 1897–98 crop results, which varied according to the quality of the land and the ability of the immigrants to work it in the face of continuous ill health. He was chiefly appalled, however, by the condition of the drinking water. Pumped from the alluvial soil or the river, it appeared clear at first and then turned red from the iron content. The company had installed filters, but this did little to make the water safe. Because of the persistent fevers, Rossati suggested the construction of artesian wells and the reclamation of the swamplands. Otherwise, he said, he could not recommend that any of the immigrants remain in such a life-threatening environment. Even with such a negative report, Ruspoli's continued influence with Italian officials prevented decisive action.[11]

With little prospect of receiving help in addressing their concerns, the immigrants began to leave Sunnyside Plantation. The local managers coerced some Italians to stay, but many of those families who had freed themselves from debt settled in towns such as St. James, Missouri, and Irondale, Alabama. Father Bandini took a group to Tontitown in northwestern Arkansas and organized what became a successful Italian enterprise on land wholly owned by the immigrants. In an effort to stem the loss of Italian workers and to avoid an 1898 crop loss, George Edgell agreed to the colonists' demands for rental contracts. This capitulation came too late, however. Between January and the end of March 1898, the population of Sunnyside shrank from 174 families to only 38. Edgell blamed the flight of the Italians on their fear of disease and regretted that so much money had been wasted in starting the colony in the first place.[12]

With the depressed cotton prices of the 1890s and the near dissolution of Austin Corbin's colonization enterprise, Italian officials lost interest in Sunnyside and other Delta plantations employing Italian labor.[13] Within a few years, however, schemes to fill the South's labor shortage with Italian immigrants revived along with the level of cotton prices. O. B. Crittenden, Leroy Percy, and Morris Rosenstock, entrepreneurs and landowners from Greenville, Mississippi, had leased Sunnyside from the Corbin heirs in 1898. Under their management, the Sunnyside Company employed agents in Italy, New York, and New Orleans to lure Italian immigrants to the plantation. Percy, a well-known politician, joined other Delta landholders in a publicity campaign to convince both their fellow southerners and Italian officials of the benefits of Italian immigration to the South. Colonization not only would solve the South's labor shortage, they insisted, but also would alleviate the problems of unemployment and poverty in the home country. In 1905 landholders and railroad and shipping executives organized a widely advertised southern tour for Edmondo Mayor Des Planches, who had replaced Francesco Saverio Fava as Italian ambassador to the United States. Des Planches made a two-day visit to Sunnyside as well as to other sites employing Italian workers. By then, 127 Italian families lived under rental contracts at Sunnyside. Although Percy and his partners lavished the ambassador with southern hospitality and assurances that the Italians were flourishing, Des Planches was much less impressed than his hosts were led to believe. Writing several years later, Des Planches pronounced that Sunnyside was nothing more than a speculative enterprise: "The company is a company of speculative venture. It tries to obtain the maximum profit from its immigrants without reciprocating on their behalf. The Italian immigrant at Sunnyside is a human production machine. He is better off than the black man, more perfect than he is, but like the black man, still a machine." Despite his pessimism, the ambassador conceded that Sunnyside might serve as a place for Italians to become acclimated to the southern climate and culture and to raise capital before moving on to better opportunities in their adopted country.[14]

An Italian farmer's home, Sunnyside Plantation, Sunnyside, Arkansas.
(Private Collection of Ernesto R. Milani.)

An Italian farmer's home, near Sunnyside, Arkansas.
(Private Collection of Ernesto R. Milani.)

Des Planches was unaware at the time of the tactics used by Sunnyside to recruit Italian immigrants—tactics that circumvented both Italian and American contract labor laws. Umberto Pierini, a Sunnyside storekeeper-turned-labor agent, sent forged affidavits from Sunnyside tenants to his subagents in Italy who used the documents to recruit the tenants' relatives for the plantation. The Sunnyside Company advanced transportation funds to those families who agreed to emigrate, and Pierini received from two to five dollars per family in addition to one dollar per acre of land that they cultivated. Pierini's commissions, along with the transportation fares and other advances, were deducted from the immigrants' first cotton receipts.[15]

The newcomers recruited by Sunnyside labor agents were little happier than their predecessors had been under the absentee Corbin landlords. Complaints from the tenants reached Lionello Scelsi and Luigi Villari, the Italian consul and vice-consul in New Orleans. Villari visited the plantation in both 1906 and 1907 and reported the now familiar set of conditions: unsanitary water, substandard housing, inadequate, expensive medical care, and rapacious prices and interest costs for goods at the company store. Moreover, the consuls learned that the colonists were afraid to leave the premises or to voice their protests openly— conditions that suggested genuine debt peonage.[16] Scelsi went further than any previous Italian official had by urging the ambassador to request that the United States undertake an investigation. At the same time, and with more immediate effect, he refused to sign any affidavit without the presence of the requesting parties. This action at once eliminated the fraudulent documents that the subagents had for so long supplied the émigrés. Leroy Percy and his partners protested, but they could not regain the favor which Count Ruspoli and Ambassador Fava had once held for the Sunnyside Company. With fresh supplies of labor withheld, Percy was hard-pressed to keep the momentum for expansion going.[17]

Consul Scelsi's complaint, which Ambassador Des Planches registered with the United States Department of State, did result in federal action. In 1907 Assistant United States Attorney Mary Grace Quackenbos was assigned to investigate violations of contract labor law by southern companies, and, at Des Planches' request, Sunnyside was placed at the top of the list. Quackenbos produced a comprehensive description of conditions at Sunnyside in her report to the attorney general and negotiated a few improvements for the immigrants with Percy and his partners, but the United States did little else in behalf of the Sunnyside tenants.[18]

Within a few years the Sunnyside experiment with Italian labor came to an end. Consul Scelsi succeeded in thwarting the recruitment of Italian families by the company's labor agents. By 1912, a boll weevil infestation and flood damage from the Mississippi River had forced many Italian immigrants from their rented Delta lands. The number of Italian families in the Delta dropped from 750 in 1907 to approximately 350 in 1912, and on Sunnyside the number fell from 128 to 60 families during the same period. By the outbreak of World War I, immigration to the South from Italy had all but ceased, and emigration bulletins made no further mention of Sunnyside.[19]

Why was the Italian government so slow to act in behalf of its citizens who had emigrated to the American South? The answer lies in a deep ambivalence at home. Italy was struggling with the massive social, political, and economic upheavals of the early stages of industrialization. To promote emigration was an admission of failed policies, but it also was one solution to the problems of overcrowding and poverty. National leadership was in the hands of the same narrow class of aristocrats that had followed Count Cavour of Sardinia in the Risorgimento. They were unlikely to take the plight of disfranchised peasants too seriously. The small Italian bureaucracy, moreover, was ill equipped to deal with the vast number of Italians emigrating to the Americas.

In the case of Sunnyside, political favoritism and the conventions of diplomacy constrained Italian officials in responding to complaints coming from the immigrants there. Sunnyside was not just a remote and unimportant plantation. Because of its size and the publicity that Austin Corbin and later Leroy Percy generated, it was viewed as a model that could be emulated not only in the American South but in Brazil and Argentina, where even greater numbers of Italian farmers had settled between the 1880s and World War I. Since Count Ruspoli and his own investments were involved in the Sunnyside experiment, the embassy and authorities in Rome were not eager to acknowledge the mistreatment of Italian citizens. In addition the ability of Italian diplomats to investigate locally was circumscribed. For the most part, consuls and vice-consuls were expected to be amiable representatives, not troublemakers in the host country. To be sure, Rossati had managed to air his reservations in the 1890s, but Father Bandini and the Sunnyside committee had reason to complain that more frequent and more thorough oversight was needed. The efforts of Des Planches, Villari, and Scelsi in the following decade produced the only practical result in the eighteen-year history of abuses of Italian immigrants: they reduced contract labor violations and made it almost impossible for Italians to settle in the southern states. Sunnyside, which closed its books in 1913 on the few remaining tenants, failed as a profitable enterprise for much the same reason that the South itself had failed in 1865: the master class refused to recognize the injustice of its labor system.

Mary Grace Quackenbos and the Federal Campaign against Peonage

Randolph H. Boehm

On June 4, 1907, the Italian ambassador to the United States wrote Secretary of State Elihu Root to complain about the treatment of Italian immigrants laboring on cotton plantations in the Mississippi Valley. Ambassador Edmondo Mayor des Planches noted that the Italian government had received "numberless complaints" from the Sunnyside colony on the Arkansas side of the Mississippi River. Des Planches had learned that Mary Grace Quackenbos was employed by the Department of Justice as a special assistant to investigate cases involving immigrant laborers held in peonage in southern states, and he requested that the American government assign Quackenbos to investigate the complaints from Italians laboring in the Mississippi Valley.[1]

Attorney General Charles Bonaparte's decision to dispatch Mary Grace Quackenbos to Mississippi could not have come at a worse time for the O. B. Crittenden Company, which leased the Sunnyside Plantation. The Crittenden Company—and especially its articulate and influential senior partner, Leroy Percy—had assumed a leading role in the locally popular vision of filling the Mississippi Delta's labor needs with Italian immigrants. In short supply of field labor and dissatisfied with the lack of discipline among black plantation hands, several large cotton plantations in the Mississippi Delta began "experimenting" with Italian immigrants at the turn of the century. The leader in the movement was the Sunnyside Plantation, a sixteen-thousand-acre tract on the Arkansas

Immigration agent in Italy.
(From Berardinelli Report, March 1909, Courtesy National Archives.)

side of the river about seven miles from Greenville, Mississippi. It was here that Italians were first colonized in the region in 1896, but the original colony nearly dissolved in the wake of a malaria epidemic and general disillusionment among the Italian colonists, and, in 1898, the property was leased to the O. B. Crittenden Company of Greenville, Mississippi.[2] The company was a cotton factorage firm that also invested in large plantations in the Greenville area. Its senior partners were Orlando B. Crittenden, a wealthy planter; Morris Rosenstock, a prominent Greenville merchant; and Leroy Percy, a lawyer, cotton planter, and Delta political leader. Despite the Sunnyside colony's inauspicious beginning, the Crittenden Company redoubled its efforts to recruit immigrants from Italy to work the plantation's cotton fields. By 1906 a stream of laudatory articles praised the experiment and suggested that Italian immigrants might prove to be the best hope for maximizing yields in the cotton kingdom of the Mississippi Delta. Contrasting the Italian renters with indigenous black plantation laborers, the commentators eagerly reported the higher average crop yields attained by the Italians and the greater care they took in maintaining the plantation lands and stock. A number of other Delta plantations imitated the Sunnyside experiment and began to recruit Italians to work their cotton fields. In 1907 Mary Grace Quackenbos reported that she had discovered forty plantations in the Delta region that worked anywhere from one to 180 Italian families.[3]

By the spring of the same year, however, the Sunnyside Plantation was beset with discontent among its immigrant tenants. Unusually heavy rains the previous autumn had diminished the yield of the 1906 crop. Letters complaining of abusive treatment by the plantation managers and unhealthy living conditions were reaching the Italian government with such frequency that the New Orleans consulate effectively closed off immigration to the entire Delta region. Only a few months before Quackenbos's arrival, a barn burning occurred on one of the subdivisions of Sunnyside, and managers attributed the act of vandalism to malcontents among the Italian tenants. Even more ominous was a rash of absconding families after advances were paid to the tenants in the spring of 1907. Leroy Percy worried that if the exodus were not stemmed, it might end in the ruin of the entire plantation.[4] In an effort to shore up morale, the company offered an adjustment in its method of marketing the tenants' cotton crop: it would allow those who had paid their debts to the company to sell the balance of their cotton themselves. Until then, the company had monopolized the sale of the entire cotton crop; they bought the cotton low from the tenants and sold high across the river in

Office of an immigration agent in a small Mississippi town.
(From Berardinelli Report, March 1909, Courtesy National Archives.)

the Greenville market, making a profit of as much as 40 percent in what was essentially a brokering role. The customary brokerage fee charged by cotton factors was only 2.5 percent.[5]

It is doubtful that the company's concessions made any difference to the more recent arrivals. Very few of the newer tenants were able to extricate themselves from debt to the Crittenden Company within the first year or two of their tenure, so they fell outside the privileged group of renters allowed to market their own cotton crops. A glance at the available immigration statistics suggests that the O. B. Crittenden Company was attempting to expand the labor force on Sunnyside more dramatically than at any time since the colony was originally founded in 1895–96. In 1900 there were approximately forty Italian families residing on the property. By the summer of 1907, there were more than 160 Italian families on Sunnyside, not counting those who had run away in the spring. Ninety-five families—approximately 60 percent of the total Italian population—were imported in 1905 and 1906 alone.[6]

The rapid increase in immigration was undoubtedly spurred by rising cotton prices after the turn of the century; however, it was accomplished largely through the systematic violation of the federal alien contract labor law, enacted in 1885, which had made it a crime to import immigrants to the United States to fill labor contracts that were made with the aliens in their native countries. The Crittenden Company's scheme to evade the federal proscription on contract laborers started in the Sunnyside Plantation store, where tenants were induced to suggest the names of friends and relatives in Italy. The company then forwarded money to ticket agents for transatlantic passage in the names of the tenants' friends or relatives. The tickets were sent to labor agents in Italy who attempted to persuade those whose names they had received to emigrate to the Sunnyside Plantation. If the entreaties proved unsuccessful, the agents simply drafted substitutes and provided them the free transportation under the aliases of the original names. The immigrants were promised jobs on Sunnyside and were supplied with a primer on answering the questions of United States immigration inspectors in order to conceal their true status.[7]

By the beginning of 1907, the ninety-five newer families would scarcely have been acclimated to the stifling heat, pestilent mosquitoes, malarial fevers, and unhealthy drinking waters common to the Mississippi Delta. The labor agents' contracts to import Sunnyside tenants were careful to conceal from prospective immigrants the environmental jolt they would experience in relocating from a Mediterranean to a semitropical climate. In addition to the climatic shock, new arrivals were further disillusioned by the company's exploitative business practices, which had also been glossed over by the importers. The forty-five-dollar per head charge for the prepaid steamship tickets from Italy were held against the accounts of the new arrivals. Their transportation debts were quickly augmented by the requirement to purchase such necessities as mules, cotton seed, tools, and living provisions from the company store at steep prices. At the close of their first crop year they would find that all of their debts were compounded by a 10 percent *flat* annual interest rate, and the low prices the company monopoly paid them for their cotton often did not liquidate the year's indebtedness.[8] The crop damage resulting from the autumn rains in 1906 further diminished whatever slim hope many of the newer families had of extricating themselves from debt to the company.

It was possible that some of these newcomers might share in the success of the established tenants who were the subjects of Delta promoters' pro-immigration propaganda, but from their own vantage point in the spring of 1907, it probably seemed far more likely that they would be overcome with illness and slip deep into debt to the O. B.

Italian family on Sunnyside that was unable to pay debts for several years because of continuing illness.
(From Berardinelli Report, March 1909, Courtesy National Archives.)

Crittenden Company, as indeed many did.[9] In its rush to lure Italians to the Delta through shady and deceptive labor importers as well as by its gouging business practices, the O. B. Crittenden Company represented more of a nightmare than an opportunity to many of the newcomers in 1907. Until Mary Grace Quackenbos challenged the company's practices that year, the rush to obtain Italian labor and expand the cotton crop took precedence over matters of fairness and legality in the plantation's operation.[10]

Mary Grace Quackenbos was the product of a radically different environment—the great metropolis of New York—where she pioneered as a woman lawyer in the immigrant slums of lower Manhattan. She had single-handledly launched the "People's Law Firm" in 1905 in order to supplement the work of the city's Legal Aid Society among immigrants, laborers, and the poor in general. This ambitious undertaking, which she subsidized in part through a modest inheritance, reveals her as very much a personification of the socially conscious "New Woman." She challenged the traditional gender roles of the nineteenth century and confidently asserted herself in the traditional male activities such as politics and the legal profession. Her political and professional audacity was heightened by an intense commitment to see that justice should not be denied to the poor. This commitment was evident in her undertaking a dangerous private investigation in 1906 of labor camps in Florida, Alabama, and Tennessee that were suspected of practicing peonage on immigrants lured from New York. On her return she presented the results of the investigation in the form of scores of depositions to the Justice Department, and she so impressed Assistant Attorney General Charles Wells Russell that he moved to have her brought into the Justice Department as a special assistant United States attorney—the first woman ever assigned to a United States attorney's office.[11]

On the surface there is more than a little irony in the scenario of an advocate for the poor from the slums of Manhattan spearheading the federal investigations of peonage in the South. The irony elicited outrage from Floridians who resented the northern woman's meddling in their labor relations; similarly, the Delta planters were outraged that a woman unfamiliar with the intricacies of cotton planting would dare challenge the standard business practices of their region.[12] Yet beneath the irony there lay a connecting thread between Quackenbos's vocation as a lawyer for the poor and her career as a federal investigator in the South. Starting from her earliest legal work in New York, her caseload was filled with complaints of laborers who had been abused or cheated by employers—employers who, in their determination to wring a profit, would scuttle fairness and even defy the law.[13] This was a problem that transcended sectional or even national differences. It was precisely the problem that led to unrest on the Sunnyside Plantation in 1906 and 1907. Quackenbos's experience with and alertness to the abuse of immigrants in New York, combined with her talents as an investigator, served her well in the South, where she picked apart the web of exploitation on a Delta cotton plantation.

Quackenbos's first strategy for investigating Sunnyside was to proceed undercover as she had in her peonage investigations in the southeastern states. The plan misfired badly, however, as one of the Italian-speaking detectives in her party was apprehended by the plantation managers, jailed overnight in the Sunnyside store, tried, and convicted of trespass by a local justice of the peace. The federal agent was sentenced to three months on the chain gang in lieu of a one-hundred-dollar fine, and Quackenbos freed him by wiring money for the fine (which he paid under protest). In Quackenbos's mind, the incident illustrated the flagrant disregard on the part of Delta planters for the elementary rights of their tenants to enjoy public access.[14]

Thwarted in her first attempt to gain access to Sunnyside and alarmed by the hostility of the plantation managers to outsiders, Quackenbos resorted to a second strategy. She approached the highest ranking sympathetic political official in the state, Acting Governor Xenophon O. Pindall, and secured his support in the form of a letter to the O. B. Crittenden Company requesting the company to grant her admission to Sunnyside. After meeting with Pindall, Quackenbos reported to Bonaparte that the governor admitted in private conversation that there were widespread violations of peonage and alien contract labor laws in the Delta region but that "politics are such that no District Attorney will care to antagonize the powerful interests of the state for the sake of these poor Italians."[15]

Armed with Pindall's letter, Quackenbos visited Leroy Percy and insisted on permission to inspect the plantation. Percy was as yet unaware of Quackenbos's earlier forray onto Sunnyside and of the agent's arrest. The overseers apparently regarded the incident as a minor matter and failed to inform the company executives upriver in Greenville of their deed. Percy was initially impressed with the woman and welcomed her proposal for an investigation. He pointed out that he had been hoping for a federal investigation for months which he thought would prove to the

Italian government that when the trouble with the tenants increased during the past spring, Percy had tried to organize a group of Delta planters to undertake a mission to Washington to request a federal investigation of Italian immigrant labor in the Delta. With Quackenbos's arrival, Percy believed that his vision of Italian plantation labor would be vindicated.[16]

Percy's openness and confidence contrast sharply with Quackenbos's initial impression of Sunnyside, influenced as it was by the complaints she heard at the Italian Consulate and the overseers' arrest of the inquiring federal agent. Indeed Percy's confidence in the probity of labor relations on Sunnyside Plantation remained steady, and apparently sincere, throughout the ensuing controversy. Even Quackenbos herself was initially impressed. "Mr. Percy appears to me to be a man of common sense," she wrote Bonaparte after her initial meeting with him. She harbored her own visions that Percy would see the sense in her proposals for improving labor relations and living conditions on several planes. He was at once, "a man of common sense," and a compelling visionary, but also a callous businessman, and an unscrupulous adversary. His ability to project the persona of the well-intentioned (later aggrieved) visionary served to confuse the situation on Sunnyside Plantation for several contemporary observers.

With Percy's permission, Quackenbos was given the run of the plantation for several days. She remained each night and rose with the Italian tenants as the morning bells rang at 5:00 A.M. She reported that she was watched by the overseers but not interfered with. After three days of investigation, she was recalled briefly to New York to give testimony in another case, but she returned within days to resume the investigation. This time, Percy chafed at her presence. Perhaps he was finally informed of the agent's arrest or warned by the overseers that her line of questioning was not what the management expected. At any rate, Percy threw obstructions in her way. He insisted that she use the company's Italian interpreter in questioning the Sunnyside Italians. Quackenbos refused, and after a standoff, Percy relented, but he then insisted she leave the plantation after each day's work. The boat trip between Greenville and Sunnyside took two hours, so Percy's requirement imposed a daily commute of four hours. Here Quackenbos was forced to yield. More significant, both Percy and O. B. Crittenden both insisted on escorting the federal investigators every minute of the day, providing what was surely an imposing presence before the Italian tenants. In spite of obstructions, Quackenbos persevered. She returned for two more days of observation and interviews, accompanied

Morning mass at Sunnyside. *(From Mary Grace Quackenbos's report to the United States Attorney General, September 27, 1907—Courtesy National Archives.)*

from the Greenville docks to Sunnyside and back by the annoyed Percy and Crittenden. The tension which the partners' presence lent to the process is apparent from Quackenbos's report that, "The Italians bravely stated their complaints before their employers in my presence," and after five days of searching investigation, she had interviewed 70 of the 157 Italian families she found on Sunnyside. She was given access to the company's books at the plantation store, where she saw incontrovertible evidence that the Crittenden Company had paid for contract laborers in violation of federal immigration laws and that it was violating state usury laws by charging the tenants "flat" interest on all advances.[17]

The tour of the plantation reinforced the negative impressions she had formed at the time of the agent's arrest and chain-gang sentence. Her interviews and her study of the Sunnyside account books disclosed widespread violations of the alien contract labor law, monopoly business practices by the company in buying and marketing tenants' cotton, price gouging on food and farm implements, substandard living conditions, serious health problems, profiteering on advances for medical care and medicines, and a pervasive atmosphere of intimidation fostered by the walking bosses.[18]

One of the most objectionable managerial practices in Quackenbos's view was the "pay roll" system. Under a clause in the tenant contract that reserved to the company the right to provide assistance in planting and harvesting the crop, company managers could assign day laborers to a tenant's fields to assist in picking the cotton. The entire amount of day-labor wages—typically one dollar a day per laborer—would be charged against the tenant's account. Quackenbos's report describes families who were plunged deep into debt because of the "pay roll," but it does not indicate how widespread the practice was. She points out that it could be used by the company to rush cotton to market during sudden price increases. However, the examples she provides describe families whose work force had been depleted by illness and death from malaria. These unfortunates would already have had their accounts run up by doctor bills, medicines, and perhaps coffin and burial costs (again compounded by 10 percent flat interest), and the added burden of paying day laborers during picking season was both a bitter humiliation and financially devastating. Even if it were applied sparingly, the "pay roll" occasioned indignant complaints from the Sunnyside tenants.[19]

On the matter of peonage, Quackenbos was initially unsure. She wrote Attorney General Bonaparte that "the bondage at Sunnyside is not exactly peonage as I understand it." The vivid evidence of physical brutality and forced labor which she found in the peon camps of the southeastern turpentine regions was not readily apparent at Sunnyside. Quackenbos's early reports indicated a more subtle system of intimidation. She noted that the tenants had no advocate on the plantation to whom they could turn in matters of dispute. Realizing their isolation and vulnerability, the Italians were subject to intimidation by verbal threats and intermittent acts of violence by plantation managers.[20]

Quackenbos wrote three increasingly authoritative reports on the Sunnyside Plantation for the Justice Department. Despite her lack of background in cotton culture and plantation management, she succeeded in dissecting every aspect of the tenants' relationship with the plantation management, and she highlighted every point of exploitation. Many of these points were matters of private contract law that were plainly outside the jurisdiction of the federal government in 1907, but they were germane to the original inquiry of the Italian government in wanting to discern the causes of the tenants' complaints.[21]

Despite her trenchant criticism of the plantation's business practices, Quackenbos held out the hope that if Sunnyside could be reformed it might serve as a model for equitable labor relations throughout the Delta. Although she expressed doubts that the immigrants should be encouraged to remain permanently in so unhealthy a location, she conferred with the Italian consulate in New Orleans and entered into negotiations with Leroy Percy and O. B. Crittenden to change the labor contract for the Sunnyside tenants. She succeeded in negotiating thirteen significant changes in the contract. Although the company rejected her suggestions to dig an artesian well, screen the cottages against mosquitoes, and lower its 10 percent interest charges on advances made to the tenants, it did agree to cease the practice of charging illegal flat interest, to allow a representative of the Italian consulate to adjudicate disputes, and to carry forth several other improvements. The most significant concession was that the company agreed to add an option to the clause that required tenants to sell their cotton outright to the company after ginning. It now would provide the option of marketing the tenants' cotton for the best price obtainable in Greenville minus the standard 2.5 percent commission. This change fell short of granting the right to free sales of cotton by the renters, but it did seem to be a major concession of a practice whereby the company allegedly reaped 30 to 40 percent profits by buying cotton low on the plantation and selling high a few miles up the river in Greenville. Percy and Crittenden evaded Quackenbos's request that the company cease the illegal importation of alien contract laborers.[22]

To justify negotiating on labor conditions, which were then clearly outside federal jurisdiction, Quackenbos claimed that she was responding to the invitation of Percy and Crittenden for her suggestions on improving the plantation. However, it is clear that she welcomed the opportunity to exert federal authority or at least federal guidance in the field of labor relations, especially as it concerned recent immigrants. She wrote the attorney general from Mississippi with "suggestions which may help solve the tangled labor problem of the South." These included drafting a standard federally sanctioned labor contract for use on cotton plantations, establishing a federal labor bureau that would distribute immigrants only to reputable employers, and collecting complaints against abusive employers and justices of the peace. Perhaps she acted too forcefully, but Mary Grace Quackenbos thought she was bringing the "square deal" to the Italian colonies in the Mississippi Delta single-handedly. She wrote Bonaparte detailing her entire initiative and asked him if she had carried matters too far in pushing for reforms in the private contract. Bonaparte responded by congratulating her on her "material progress" and then wrote the President pointing out that the woman's work in the Delta would be to "good moral effect" even if it did not turn up any clear evidence of peonage.[23]

Quackenbos was sufficiently impressed with the progress she thought she was making with Percy that she accepted a dinner invitation the planter graciously extended to her. During her social evening with the Percy's Quackenbos sensed the possibility of establishing feminine bond with Leroy's wife, Camille, and the New Yorker sought to interest the plantation mistress in the welfare of the Sunnyside Italians. Camille assured her guest that she would maintain regular liaison with the Italian colony, providing the woman's touch that Quackenbos found wanting in the management of the plantation. Shortly after the dinner, however, the Italian priest on Sunnyside sought out Quackenbos to report that Camille Percy complained bitterly about her to the cleric, threatening to have the investigator arrested if she dared set foot on the plantation again without Leroy's permission. At about the same time an Italian family also slipped a note to Quackenbos begging for her to return to the plantation to adjudicate a dispute with the plantation store. Sensing betrayal, the investigator redoubled her investigation of contract labor and peonage investigations.[24]

As Quackenbos had noted in her very first communication from Mississippi to the Attorney General, although an atmosphere of intimidation was readily apparent on the plantation, evidence of forced labor was not. Sensing that it would be difficult if not impossible to elicit evidence of peonage from the tenants with the company partners at her elbow, Quackenbos boldly requested that she be provided with several dozen blank subpoenas to use as necessary in calling prospective witnesses from the plantation to the U.S. Attorney's office in Jackson where she and her interpreter could question them without interference from the planters and overseers. The request was dismissed as unorthodox and inappropriate, but again she tenaciously persevered with the investigation. During her earlier work inspecting living conditions on the plantation, she heard evidence about families who had fallen into peonage but since been allowed to leave the plantation. As she pursued the lead, she developed evidence that peonage had been resorted to by the Sunnyside overseers, but not with regularity. She learned that almost thirty families had left Sunnyside owing debts. The large majority of abscondees were allowed to leave or at least they were not pursued. However, in a few cases, absconding tenants were pursued and forced back to the plantation, sometimes with threats of being sentenced to the chain gang if they resisted. In what would become the most notorious case, two men, having fallen deep into debt to the company, decided to

Italian tenant who escaped a Mississippi plantation with her husband to work in Alabama. The oven in background, at the immigrant's home in Alabama, was used for baking bread.

(From Berardinelli Report, March 1909, Courtesy National Archives.)

leave their families to work the Sunnyside crop while they went to find jobs in the mines in Alabama. They reported their intentions to O. B. Crittenden in his office in Greenville, but before the train left the station, Crittenden summoned a local sheriff, and together they pulled the Italians from the train and forced them back to Sunnyside. Although Crittenden denied the Italians' contention that he and the sheriff used physical force, he argued that he threatened the two with prosecution under the Mississippi state statute that made it a crime to leave an employer while in debt. A month later, the two peons and their families were roughly evicted from the plantation. They were cast upon the levee at night, where they remained without shelter with an infant until the launch arrived the next morning. They ultimately made their way to Birmingham, Alabama, where Quackenbos traced them on information given by Sunnyside tenants. Working with the attorney general's office in Washington and with the local United States attorney in Jackson, Mississippi, Quackenbos drew a petition to indict O. B. Crittenden himself for peonage.[25]

Percy learned that Quackenbos's investigation had shifted from living and working conditions on Sunnyside to criminal violations of federal peonage laws weeks before Crittenden was charged. Shortly after

Quackenbos completed her initial investigation and her negotiations with Percy and Crittenden, Percy wrote Attorney General Bonaparte to complain about Quackenbos's biased disposition in favor of the less-prosperous tenants on the plantation: "She takes no interest whatever in the laborer who is prospering. It is the laborer who is unfortunate or sick who appeals to her womanly sympathies, and on whom she concentrates her attention." Bonaparte replied that Quackenbos's only mission was to discern whether peonage existed. "The question is not whether any particular 'colonists' are, or are not, prosperous, but whether they are, or are not, free, as the law requires them to be." Curiously, Bonaparte did not mention that the wider range of Quackenbos's investigation was consonant with the original request of the Italian ambassador.[26]

Shortly after the attorney general divulged the fact that Quackenbos was investigating peonage on Sunnyside, the local Delta press began a campaign of vilification against her. The Vicksburg *Herald*, the Delta's leading daily, edited by Captain J. S. McNeily, a Percy relative and political ally, announced:

> It has come to The Herald's knowledge that Attorney General Bonaparte is co-operating with the Italian consul at New Orleans in breaking up immigration to the lower Delta. The agents of his department

Italian children who were expelled from school in Sumrall, Mississippi, because they were considered "inferior."
(From Berardinelli Report, March 1909, Courtesy National Archives.)

have visited these immigrants on the plantations—taking advantage of their ignorance and suspicious natures to excite discontent and hostility toward their employers. . . . This is a remarkable accusation to lodge, but it is warranted by the facts of the labors of a woman lawyer, Mrs. Mary Grace Quackenbos; Special Assistant U.S. Attorney and Socialist agitator.

The Greenville *Times* alerted its readers to ". . . a lady lawyer who is stirring up the Italian immigration questions from center to circumference. . . . In other words she is just raising ———, as is usual with a professional woman."[27]

Local opposition to the investigation took a strange turn when Quackenbos's legal portfolio inexplicably disappeared from her Greenville hotel—and turned up in the hands of Percy's close friend and political ally, the former United States congressman from the Delta, Thomas C. Catchings. The portfolio would have provided evidence to Mississippians that the woman was developing a credible peonage case. The campaign of obstruction against Quackenbos was intensified. In a letter to the Vicksburg *Herald*, Leroy Percy charged Quackenbos with "interfering with and demoralizing the labor, both white and black" in the Delta. The latter statement, touching as it did on the explosive subject of race, was well calculated to prejudice potential jurors should a case be brought. The loss of the legal portfolio had other serious consequences for the government's case. Quackenbos reported to Washington that her witnesses were being intimidated on Sunnyside and that the United States attorney in Jackson—yet another personal friend of Leroy Percy—had suddenly begun to express reservations about whether a prominent Delta businessman defended by Leroy Percy could be prosecuted successfully in Mississippi.[28]

The likely theft of her papers combined with the negative press campaign and the intimidation of Italian witnesses on the plantation served only to convince Quackenbos that the Sunnyside managers would resort to extralegal means to frustrate the authority of the federal government. She had in the meanwhile succeeded in eliciting a confession from the company's labor agent—the former manager of the Sunnyside store—concerning the Crittenden Company's ongoing scheme to evade the federal alien contract labor laws. She grew determined that the Delta planters should be accorded an example of federal resolve. "Only prosecutions . . . will improve matters," she declared in a letter to Bonaparte.[29]

Leroy Percy seemed confident that no Delta grand jury would indict his prominent business partner and client. He was concerned, however,

An Italian woman who was among immigrants who fled from their homes in Sumrall, Mississippi, because of threats from local townspeople.

(From Berardinelli Report, March 1909, Courtesy National Archives.)

that Quackenbos might draft additional indictments against the Sunnyside Plantation managers for other acts of forcing Italians back to the plantation.[30] These men would be more difficult to defend than his patrician partner, and their indictments might be drawn in Arkansas, where his influence was less pervasive. Even more troubling was that the case was threatening to become a public relations disaster. The *Chicago Tribune* caught wind of the investigation and ran a scandalous article about the new "slavery" in Mississippi [*sic*]. Percy feared that the bad publicity would rather prejudice the Italian government and spread back to Europe through the emigrant grapevine. If so, his vision of reopening the Delta to immigration would be in extreme jeopardy. As a final stratagem, he moved to have the Department of Justice recall Quackenbos from the South before she could wreak further havoc.

Toward this objective, he drew his trump card: his personal relationship with Theodore Roosevelt. The two had become friendly during the famous 1902 bear hunt on Smedes Plantation where Percy formed part of the Mississippi escort. The friendship was renewed by Percy's periodic visits to the White House and had only a few weeks earlier been rekindled during the president's visit to Vicksburg as part of a speaking tour through the South. Undoubtedly, Percy represented more to Roosevelt than a charming hunting companion. He was a sterling example of the New South leader—a tireless booster and promoter of southern economic development. By making his case to Roosevelt, Percy would force the president to choose between the "New South" and the "new woman."

Leroy Percy's meeting with Roosevelt in Washington was a masterpiece of advocacy. He was careful to emphasize that he had no intention of prejudicing the Crittenden case. With good reason, he was confident the Mississippi judicial system would handle that matter. Instead, he framed his argument on a fear that Quackenbos's confidential reports would be published. He argued that publication of her charges would completely undermine efforts to induce immigration to the Delta, and this in turn would threaten the economic future of the entire region. Having established his patriotic motives to the president, he proceeded to his main objective: he attacked Quackenbos's qualifications as an investigator of southern conditions and implored the president to sponsor another investigation made by "*men* of practical understanding," [emphasis is Percy's] an investigation that he believed would refute her reports of mistreatment. In the end, the president assured his friend that Quackenbos would be withdrawn from the South.[31]

Two points in Percy's argument seem to have been crucial in winning Roosevelt over, or at least in supplying the pretext Roosevelt needed to come to the aid of a personal friend caught in an awkward position by a relentless federal investigator. First, Percy twisted Quackenbos's emotional indignation—which was readily apparent in her reports—into an anti-feminist argument that she lacked judgment: "She is a philanthropic humanitarian, a doctrinaire, seeking to remove poverty wherever she finds it at the expense of others and without discrimination as to whether that poverty is due to unjust treatment or oppression, or is the result of necessary conditions and environment. . . . Her manner was that of a 'Lady Bountiful' dispensing alms." Second, Percy stressed the sectional differences between the northern women and the Delta environment, asserting that "there was not a condition, a custom, a form of contract, or a crop raised on a plantation about which she had the slightest information."[32] The president echoed both of these points in the aftermath of the affair in explaining his position on the case. In a letter to historian Albert Bushnell Hart, Roosevelt wrote:

> I have been very uneasy about Mrs. Quackenbos. She comes in the large class of people who to a genuine desire to eradicate wrong add an unsoundness of judgment which is both hysterical and sentimental. . . . The fact is that on those southern plantations we are faced with a condition of things that is very puzzling. Infamous outrages are perpetrated—outrages that would warrant radical action if they took place in Oyster Bay or Cambridge; but where they actually do occur, the surroundings, the habits of life, the sentiments of the people, are so absolutely different that we are in reality living in a different age, and we simply have to take this into account in endeavoring to enforce laws which can not be enforced save by juries. . . . It is like trying to enforce a prohibition law in New York City.[33]

Informed of Percy's criticisms by Bonaparte, Quackenbos argued that the cause of immigration to the South was best served by demonstrating that the federal government would protect the rights of immigrants in the region. She defended her reports, pointing out that she fully anticipated that every contention she made would be met with strenuous condemnation by the Crittenden Company. With that in mind, she insisted, she wrote only what could be borne out amply by the evidence. In addition, she was able to report progress in developing a coalition among federal officials and progressive planters in the Mississippi Valley who were eager to have the government proceed against peonage in the region.[34]

Her protests were unavailing. She was recalled to Washington where she drafted her final report on the conditions at Sunnyside and in the Italian colonies throughout the Delta. Her final report on Mississippi was sent to the Italian government with the reminder that it was to remain completely confidential. Despite its numerous accounts of peonage and violations of alien contract labor law, the Department of Justice did not even keep a copy for its own files.[35]

Roosevelt's sensitivity to sectional prerogatives and southern sentiment in the Sunnyside case may have been a sincere change of heart, but his explanation to Hart seems more likely to have been disingenuous. For almost the entire year previous, Quackenbos had incurred the public wrath of southern turpentine operators and their representatives in the United States Congress. Despite the efforts of the Florida Congressional delegation to have Quackenbos removed, the administration stood behind her. Indeed, as Quackenbos was investigating Sunnyside, the president clipped a report (erroneous as it turns out) that the woman lawyer was assisting the Justice Department in the prosecution of the southern timber trust, and he confided to the attorney general that he found the story "very amusing." When, in Roosevelt's next exchange with Bonaparte, he was apprised of the fact that Quackenbos was investigating his friend Leroy Percy, the amusement ceased, and Roosevelt tersely inquired of his attorney general how long it would be necessary to retain the special assistant. Since there was nearly a year in which the president could have voiced doubts about Quackenbos's "hysterical" behavior, the point at which he did begin to criticize her can only raise suspicions.[36]

Leroy Percy succeeded in preventing even so much as an indictment of O. B. Crittenden. Crittenden admitted to the Court that he threatened the Italians with arrest if they did not abandon their exodus and return to work out their debts on Sunnyside. But, a categorical denial of a planter's right to run down absconding tenants was more than the Delta jurors cared to make. In spite of the charge made to the grand jury by Federal District Judge H. C. Niles that virtually instructed them to indict, the Vicksburg panel refused to find a true bill. Judge Niles instructed the jurors that the Mississippi contract labor statute under which Crittenden forced the laborers to return "is an attempt to establish what the Constitution and law of the United States forbid, and is therefore void. . . . A State law calculated to compel a man, by threat of imprisonment, to remain in servitude, and making him *prima facie* guilty of a crime upon evidence showing only a breach of contract, violates the letter and spirit of the Constitution." In an effort to save face, the panel reported that "we feel that the thorough investigation by the Government will have a tendency to prevent peonage ever existing in our district." Assistant Attorney General Charles Wells Russell, sent to supervise the prosecution in lieu of the beleaguered Mary Grace Quackenbos, reasoned that the prosecution probably would have a salutary impact on labor relations in the Delta despite a failure to indict Crittenden or overturn the Mississippi statute facilitating peonage.[37]

Mary Grace Quackenbos never returned to the South, despite the entreaties of the assistant United States attorney in New Orleans that she establish a headquarters there from which to monitor compliance with anti-peonage laws. Her final year in the federal service ended with another disappointment as defendants in a major Florida peonage case were acquitted under circumstances at least as disconcerting as those in the Crittenden case. She returned to private practice at People's Law Firm.

The decision in the Crittenden case did end the controversy over labor relations on Sunnyside Plantation. Leroy Percy and southern representatives in Congress seized upon an opportunity to mount a counterattack on Mary Grace Quackenbos and the Department of Justice with the hope that they could blot the woman's investigation from the historical record. The opportunity presented itself in the person of Albert Bushnell Hart, professor of American history at Harvard University. While touring the South researching his work in progress, *The Southern South,* Hart was hosted by Prof. Franklin L. Riley of the University of Mississippi. Riley took Hart to Greenville and introduced him to Leroy Percy. There, Hart was inculcated with Percy's side of the Sunnyside controversy and was encouraged to make a personal inspection of the plantation. After a day's visit to Sunnyside, Albert Bushnell Hart came away with a very favorable impression. Moved by Percy's complaints that the plantation had been unjustly maligned by the federal government, Hart determined to convey his impressions to his college classmate and lifelong friend, Theodore Roosevelt. Hart wrote the President: "As to the conditons of the people on Sunnyside, both Italians and negroes [*sic*], they are better housed and better cared for than on any plantation that I have seen or visited. There is no want among them and has been none. . . . If any report has been made that the people are badly treated by the owners of the plantation I feel sure that it is erroneous. . . . As for restraint or violence on the plantation, it is on the face of it very unlikely."

Hart brushed aside the arrest of the federal agent during Quackenbos's first probe of Sunnyside by relating the overseer's alibi that the agent "behaved in a suspicious manner and absolutely declined to show any credentials other than a printed card." So far as the professor could tell, Sunnyside was "easily accessible . . . and people can come and go to it freely." Like Quackenbos, Hart was given access to the plantation record books. Where the female attorney spotted evidence of illegal payments for alien contract laborers and accounts inflated by usurious "flat" interest charges, the Harvard professor only commended the plantation on its diligent record keeping. It was true, he pointed out, that many plantations of the Delta "plundered" their hands, "Sunnyside is not a plantation of that kind." He concluded his assessment by pointing to the fact that nearly all of the Italian tenants had renewed their contracts for 1908 as proof that "they are satisfied." In a postscript, Hart qualified his praise slightly, noting "My belief is that the bottom trouble at Sunnyside is that like most of the Delta it is very subject to Malaria."[38]

While undoubtedly sincere, Albert Bushnell Hart's impressions do not stand up when compared to Quackenbos's lengthy investigation and reports. Had Hart reflected longer upon the implications of malaria among the tenants, for example, he might have grasped the chain of events that brought many of the tenant families to the state of desperation that Quackenbos found six months earlier. As malaria depleted family work forces, tenants were threatened with the prospect of carrying the "pay roll" of day laborers, as well as shouldering charges for doctors' bills, medicines, and perhaps coffin and burial charges—all compounded by 10 percent flat interest. These were circumstances, as Quackenbos reported, that could and did push tenants into dire financial straits.

Hart's failure to see the implications of malarial infection upon the tenants' financial standing can be attributed to the timing of his visit. He spent a day on the plantation during the slack and nonmalarial month of January. By contrast, Quackenbos spent five days on Sunnyside during the summer and returned, contrary to Leroy Percy's wishes, during harvest time in October, at the height of the malaria season.

Entirely apart from his failure to explore the implications of malaria on Sunnyside, his oversight of so many other matters that Quackenbos raised in her investigation casts even more serious qualifications upon Hart's impressions. He mentions nothing about the problems of water purification, screening the cottages from mosquitoes, or the autocratic control that the company exercised over the tenants' production. He either failed to discover or failed to report that there had been a rash of abscondences during the previous year. Hart and his host Riley probably relied upon visual observations and made no effort to interrogate the Italians themselves. If they did make inquiries among the tenants, presumably they would have relied upon the plantation's interpreter, whose presence seemed certain to quell any gestures of discontent among the tenants.

In spite of the ill-timed and superficial nature of his investigation, one aspect of Hart's report might seem to weigh heavily against Quackenbos's point of view: his claim that "more than 19/20 of the Italian families . . . have contracted again for 1908." However, this too was an observation that begged closer investigation. There is evidence that at the turn of 1908 the tenants might have had reason to hope that things were improving on the plantation. For all of its problems, the year 1907 seems to have ended with a bumper crop. Percy wrote Governor Pindall to brag about the tenants' windfall for 1907, ". . . the Italians, usually prosperous, have been most unusually so this year, and most of them have insisted upon contracting for another year."[39] A bumper crop would have given the Crittenden Company the wherewithal to bolster the tenants' accounts; the glare of federal attention would have provided the company with every incentive to do so. Many of the tenants probably deserved to reap windfall returns from diligent cultivation and frenzied picking during the previous year. Even for those on the margins, however, the company held it in its power to tilt the scales a bit in the tenants' favor during this crucial settlement. The points of arbitrary authority, which Mary Grace Quackenbos highlighted in her reports and which principally consisted of the company's unilateral right to weigh and grade the cotton—a privilege that could be used to squeeze tenants in poor crop years—could just as easily be used to placate tenants in a situation such as the one that existed after the glare of attention by the federal government. As with his failure to appreciate the implications of malaria, the timing of Hart's visit certainly influenced his assessment of the Italians' satisfaction. For he visited just after the settlement at the most prosperous point in the cycle of the cotton economy.

The possibility also remains that the reforms Quackenbos herself negotiated with the Crittenden Company had been honored and that these provided a measure of satisfaction to the Italian tenants. Whether Quackenbos's reforms were faithfully instituted is an issue that unfortunately remains undocumented, except in one instance. On the matter of the usurious flat interest charges, an exchange between Leroy Percy and one of the plantation's overseers shows that the Crittenden Company tried to ignore the change it had negotiated with Quackenbos and continued charging illegal flat interest. Having been alerted to the abuse by the woman lawyer, however, the Italians protested the charges. In face of the tenants' protest and the knowledge that they were breaking the law, the company finally relented, converting interest charges to a per annum basis.[40] In the absence of fuller documentation, one can only wonder how often this sort of negotiation occurred between the Italians and the plantation overseers in the weeks before Hart's arrival.

Whether the rest of Quackenbos's contract amendments were honored might in itself have explained the contented disposition Hart claimed to find. The most important of Quackenbos's amendments could have made a substantial difference in the tenants' attitude. Free quinine (sent from the Italian consulate) was supposed to have been made available to the Italians and that might have allayed the fear of malaria. The option of having their crops brokered in the Greenville market instead of surrendering their cotton to the Crittenden Company might have brought significantly higher returns to tenants. These were matters that begged for the sort of thorough investigation that might have been made by Quackenbos herself, had she been permitted to continue with her assignment to the Delta. Under Hart's casual gaze, the issue of how long the Italians had been "satisfied" and precisely why, was allowed to pass unexamined.

A final discordant note in Hart's report to the president concerns the matter of access to the plantation. Unbeknown to Professor Hart was something that should have been known to the president—that Leroy Percy had vigorously discouraged federal investigators from visiting the plantation throughout the previous autumn. He threatened Mary Grace Quakenbos with arrest when she returned for a follow-up inspection in the middle of October, during the harvest and malarial season. A few weeks later in early November, U.S. Immigration Inspector John Gruenberg was refused admission to the plantation when he insisted on using one of Quackenbos's Italian interpreters instead of an interpreter provided by the Crittenden Company. In view of Percy's determined opposition to admitting the interpreter to Sunnyside, Gruenberg reported to Washington " . . . it would be unsafe for his life to go there."[41] Albert Bushnell Hart's confident assertions about free access to Sunnyside stand in sorry contrast to the reality that access was tightly controlled by the Crittenden Company and its overseers.

In spite of its fatal flaws, Hart's report struck a responsive cord with the president. "Your letter is of real importance. I have sent it at once to Bonaparte," Roosevelt responded, tacitly admitting that it was only with Hart's letter that he was completely convinced of the merits of Percy's complaint against Quackenbos. The mass of conclusive evidence compiled by Quackenbos and Immigration Inspector Gruenberg against the

Crittenden Company for violations of peonage and immigration laws wilted before Roosevelt's trust in the impressions of his college friend. In a sad twist of irony, only a few days later, O. B. Crittenden admitted to the Federal Grand Jury at Vicksburg that he had in fact run the two Italian tenants back to the plantation from the Greenville train station in 1907.

A week after the grand jury dissolved, the U.S. attorney for the Eastern District of Arkansas, sensing the miscarriage of justice at Vicksburg, wrote Bonaparte that indictments could likely be found against the Crittenden Company by a Federal Grand Jury in Arkansas "if warranted by the facts." The suggestion fell upon deaf ears; the Justice Department had washed its hands of the case, apparently satisfied with the "moral effect" of its effort. When the Bureau of Immigration forwarded evidence obtained by Inspector John Gruenberg showing systematic violations of federal immigration laws by the Crittenden Company and two other Delta plantation companies, Bonaparte sat on it for six months. The incriminating evidence was finally forwarded to the U.S. attorney in Jackson, Robert C. Lee, a Percy crony. There, the immigration law cases quietly died, unprosecuted.[42]

The impact of Hart's report extended beyond the White House and Department of Justice. Leroy Percy obtained a copy of the letter and forwarded it to Arkansas Governor Pindall. Pindall, in turn, had the letter published in the Arkansas Gazette. The Vicksburg Herald, eager to vindicate the Crittenden Company, reprinted the letter as published in the Gazette. Bushnell Hart himself expanded upon his report in an article for the Boston Evening Transcript, attacking Quackenbos's investigation and lavishing praise on Sunnyside Plantation. With Hart's impressions out in the open, Mississippi Congressmen John Sharp Williams and Benjamin G. Humphries swung into action in the United States Congress. The two Mississippians, both representing parts of the Delta, and both warm personal friends of Leroy Percy, joined with Rep. Frank Clark of Florida in pummeling Quackenbos on the floor of the House of Representatives. Clark, who represented the Jacksonville district, which had been one of the focal points of Quackenbos's earlier prosecutions, vented his longstanding consternation at both Quackenbos and Attorney General Bonaparte, whom he accused of using peonage as a pretext to "regulate sociological conditions" in the southern states:

> It seems that in the course of his undertaking this transplanted
> bud of alleged French nobility [Bonaparte] became acquainted with a

lady bearing the euphonious cognomen of Mrs. Mary Grace Quackenbos, whose field of labor previous to her acquaintance with the great Baltimore lawyer was in the slums of the "East Side" of "dear old Manhattan Isle." I presume that the learned Mr. Bonaparte, the head of the Department of Justice, was of the opinion that this "slum worker" from New York was a very proper and fit person to send into the State of Florida and other benighted regions of the country to regulate conditions, legal and otherwise. . . .

Benjamin Humphries of Greenville, Mississippi, rehashed the charge made earlier in the Delta press that Quackenbos was a socialist: " . . . the fact is that when she undertook this investigation she was thoroughly saturated with those ideas of the proper relations of the rich and the poor, of society to man, of the individual to society, which are the high tenants of socialism." Humphries read Hart's letter to Roosevelt into the record as a final slap at Quackenbos.

Frank Clark submitted a motion to the House to set up a congressional investigation of Quackenbos's role in the peonage investigations. It was referred to the House Rules Committee, where John Sharp Williams amended the proposal. Williams moved to have the entire question of peonage investigated by the U.S. Commission on Immigration that was established by an act of Congress a year earlier to make a wide-ranging investigation and to report on immigration to America. In a significant twist, Williams amended his own motion and charged the Immigration Commission to study peonage not only in the South, but throughout the entire country. In this form, the motion passed a House floor vote.[43]

Williams' amendments proved to be deft maneuvers in smothering the voice of Mary Grace Quackenbos. Had the House approved Frank Clark's initial motion for a congressional investigation of the peonage investigations, it certainly would have had to call the forceful woman to public testimony. Given the opportunity of addressing Hart's report and her congressional critics, she would have had the opportunity to defend her reports and the record of her activities. The Immigration Commission, on the other hand, was not charged to hold public hearings. Its investigation and the drafting of its final report were off the public record. Quackenbos would not even have needed to be consulted, much less given a public hearing. Williams surely took stock, also, in the fact that one of the Senate representatives to the commission was the senior Senator from Mississippi, Anselm J. McLaurin, who might be counted upon to take a special interest in the section of the commission's work dealing with peonage.

Williams' maneuver to get the investigation to the Immigration Commission proved to be more fortuitous than he could have imagined. Anselm McLaurin died in office as the commission's work was in progress, and he was succeeded in the U.S. Senate by none other than Leroy Percy. Without a modicum of concern for a conflict of interest, Percy was appointed to McLaurin's position on the Immigration Commission. The historical record of the commission's work—the background correspondence, staff assignments, and draft reports—have apparently been lost, making it impossible to know what role Percy actually played in the investigation and draft report on peonage. However, the commission's published report must have pleased him greatly.

If critics seriously waited for the commission to sort fact from assertion about Sunnyside Plantation, they waited in vain. Sunnyside, which prompted the commission's peonage investigation, was not even mentioned in the peonage report. The content of the slender seven-page document was weighted down by the proviso, insisted upon by John Sharp Williams, that it not only report on peonage in the South but in the entire United States. Manifestly a cobble work of political compromise between northern and southern commission members, the report pays lip service to a couple of major peonage cases in the South but qualifies even those tepid allegations: "This committee found and reports that instances of peonage as above described had occurred in 1906 and 1907 in some southern States, but these were only sporadic instances, and the Commission found no general system of peonage anywhere." In an apparent effort to balance its gentle reprimand to the South, the report continues, ". . . in every State except Oklahoma and Connecticut the investigator found evidence of practices between employer and employee which . . . would constitute peonage as the Supreme Court defines it." And, "Since the evils of involuntary servitude have been largely stamped out in the southern States, there has probably existed in Maine the most complete system of peonage in the entire country." In a final flight from reality, the report asserts, "In the opinion of the Commission the vigorous prosecutions have broken up whatever tendency there was toward peonage in connection with aliens in the southern States, and the fact that juries in those States will convict even in cases of technical peonage unaccompanied by brutality would seem to indicate that offenses against alien laborers will not be permitted to go unpunished."[44]

Manuscript census returns for 1910 seem to indicate that the Italian population of Sunnyside remained much at the same level as at the end of 1907. Through whatever means or fortune—abundant crops, fair treatment, or subtle coercion—the Crittenden Company managed to retain most, and perhaps all, of its 1907 labor force on the plantation. However, it was unable to augment the immigrant labor force. If the American government had been persuaded to gloss over the record of past abuses, the Italian government was not so disposed. The census records show that only two Italian families moved to Sunnyside after 1907. The planters' dreams of populating the Delta with Italian laborers had received a mortal blow. The Crittenden Company sold its interest in the Sunnyside property after the flood of 1912 practically wiped out the plantation. The new landlord's determination to change from a rental to a sharecrop system was bitterly resisted by the Italians, and most of the colony dissolved because of their refusal to work for shares.[45]

Italian family on Sunnyside Plantation.
(From Berardinelli Report, March 1909, Courtesy National Archives.)

Peonage continued to flourish in the South in spite of the efforts of the Theodore Roosevelt administration to keep it in check. However, cases involving immigrants are unknown after 1907. To Mary Grace Quackenbos belongs a large share of the credit for that limited accomplishment. Her efforts in the Sunnyside case were closely monitored by the Italian government, and the case played a major role in curtailing Italian immigration to the lower Mississippi Valley. Quackenbos's record of courage, ability, and tenacity in the peonage investigations has fallen into relative obscurity from where it deserves to be rescued. Equally worthy of rescue are her numerous investigative reports of plantation labor conditions in the New South and of the evasion of federal immigration laws. Quackenbos's most significant legacy may well lie in the wealth of her observations and her diligence in recording them.[46]

Leroy Percy and Sunnyside: Planter Mentality and Italian Peonage in the Mississippi Delta

Bertram Wyatt-Brown

Over Leroy Percy's grave in the tree-shaded Greenville cemetery there stands Malvina Hoffman's bronze knight, allegedly emblematic of the occupier upon whose dust the melancholy guardian fixes his downcast eyes. Both Leroy's son, Will Percy, in his classic *Lanterns on the Levee*, and Lewis Baker, the family's most recent chronicler, considered Percy a stainless and indomitable leader of his section and Delta homeland. Baker stoutly vindicates Percy's operation of the eleven-thousand-acre Sunnyside Plantation on three grounds: first, Percy had a legal rental contract; second, Mary Grace Quackenbos, his federal nemesis, was prejudiced against him; and third, Percy was a shining idealist.[1] Baker's argument needs anything but reaffirmation. Rather, Percy's handling of the Sunnyside controversy must be placed in its cultural context. Such an undertaking demonstrates not just the persistence of an old planter mentality but also a record of momentary success and ultimate failure for all concerned in the Sunnyside operation.

Leroy Percy's region was a victim of its own past and therefore of a continuing, intractable problem: the centuries-old heritage of slavery that had implanted in every white man habits and expectations not quickly broken, and had implanted as well the reasons that had prompted the use of coerced labor. Abundance of land and insufficient laborers to cultivate it had led to primitive and unsubtle methods of settlement throughout southern history: indentured servitude, slavery, and finally post-emancipation debt peonage for both blacks and whites. As Delta planters saw it, the labor situation had reached a crisis when cotton prices recovered from their disastrous plunge throughout the 1890s. New lands could not be opened up without a replenished labor pool. Charles Scott of Rosedale, Mississippi, owner of thirty thousand Delta acres and a friend of Percy, wrote to Stuyvesant Fish of the Illinois Central Railroad, Greenville's link to the world: "The labor situation here is worse than I have ever known it, and unless we can get some additional labor, considerable quantities of land in the Delta will lie fallow next season."[2]

Sharing Scott's concerns, Leroy Percy was a product of the old style of southern leadership, but he hoped to capitalize on a predecessor's initiative. Sunnyside Plantation rimmed the sickle-shaped Lake Chicot in Arkansas, across the Mississippi from Greenville, Percy's home. In 1895 Austin Corbin, a Chicago industrialist and philanthropist, had begun the experiment of corporate southern agriculture with the use of Italian labor. Corbin's death, low cotton prices during the 1890s depression, and a yellow fever epidemic led to its failure. In 1898 Percy, O. B. Crittenden, a cotton broker, and Morris Rosenstock, one of Greenville's several prominent Jewish entrepreneurs, seized the opportunity and began anew. They leased the land from Corbin's heirs and, like Corbin himself, imported northern Italians through labor agents in New York and New Orleans. Year by year, the community grew. Owners and renters alike benefited from rising demand for cotton. With his political connections and a flair for publicity, Percy widely advertised the company's purported success. Other planters in the region imitated the Percy scheme. Like him, they claimed to prefer European peasants over black laborers.[3] By 1907, Percy had managed to settle about 158 families on the plantation.

Soon enough, however, criticisms arose. Disillusioned settlers wrote home about the poor health conditions, the indifference of the owners, the contempt in which they were held by native whites, and the terms of their contracts, which, as they saw it, reduced them to a state of debt peonage. In response, in 1905 Ambassador Edmondo Mayor des Planches toured Italian settlements, including Sunnyside, which he found much less impressive than Percy's campaign in the public press proclaimed.[4]

Under the leadership of Des Planches, Italian authorities in Rome, Washington, and New Orleans began to throw obstacles in the path of Delta employers. Prodded by their animadversions, the Roosevelt administration turned its attention to issues of suspected peonage and violations of the alien contract labor laws, which, since 1885, had had the support of organized labor. In the early summer of 1907, with peonage cases already pending against timber, turpentine, citrus, and mining interests elsewhere in the lower South, Henry L. Stimson and General Charles W. Russell of the United States Attorney General's staff authorized Mary Grace Quackenbos, a special assistant with experience in the labor field, to investigate the situation at Sunnyside and other Delta plantations.

Over the next few months, Quackenbos, appalled by the conditions she saw, and Percy, equally dismayed by having to deal with so spirited a reincarnation of an abolitionist "bluestocking," became sworn enemies. Although many of her findings remained buried in comprehensive reports from the field to Justice Department officials, she apprised the public of her Delta work with the same shrewdness and sense of timing that Leroy Percy had employed to defend himself and the Arkansas plantation. She accused him (as well as many others) of cheating the helpless Italians and thereby of ruining the South's opportunity to recruit effective workers from abroad.[5]

The list of the company's offenses that she drew up was formidable: she accused the company of enticing the Italian immigrants with false promises and supplying them with fraudulent immigration papers; of overcharging them for the rented lands, which were sometimes unimproved or marginally productive, and for goods purchased at a monopolistic company store; of providing them with unclean water sources, unscreened and leaking cabins, and an inadequate and overly expensive medical service; of forcing them to pay exorbitant transportation and ginning fees; and finally of undervaluing the cotton that they could sell only through the Sunnyside Company. Worst of all, strangers in an alien countryside, the Italians were easily intimidated into docility. With none to help them, they had to remain on the land or risk apprehension by Percy's armed and uninhibited managers and corrupt law enforcement officials.[6]

The combination of Italian and federal skepticism and inquiry was indeed daunting. Yet, Leroy Percy could make a legitimate claim against charges of brutal exploitation. A single word suffices: profit. The rental system used at Sunnyside worked to the Italian colonists' advantage—so he and his partners Crittenden and Rosenstock insisted with apparent justice. The Sunnyside lands—at least those acres that were properly drained—were extraordinarily fertile, even by Delta standards. Often Percy explained that many of the colonists had done exceedingly well.[7] Writing to James Watkins in New York City, he observed in early 1908, "Some of the Italians working no more than 40 acres of land have saved up in the last eight or nine years, as much as $15,000.00 in cash." A man and wife, he wrote, could raise a cotton crop of twelve bales on only fifteen acres. At $20.00 a bale (500 pounds at about twelve cents a pound), the return came to $720. Half the amount belonged to the couple. $360 was not a grand sum, Percy conceded. But the settler "has the advantage of living in the country and raising a crop, and he can, out of this, save money. If he is able to rent the land, he can do it for one-fourth of the crop, or at an average rental of about $7.00 per acre. This would leave him, under the same estimate, $605.00 as the result. There is no trouble about his securing day work, at fair wages," Percy continued, "for all of the time he is not in his own crop, so as to help defray his living expenses."[8]

More trustworthy testimony appeared at the height of the controversy. On a visit to Sunnyside in early 1908, Albert Bushnell Hart, a professor of history at Harvard, examined peonage charges against the firm in behalf of President Theodore Roosevelt, his close friend. Hart claimed that 95 percent of the families had renewed their contracts in 1907 for the next season. A former accountant, Hart looked over the company books and pronounced them "comprehensive and honest." Hart further observed that he "saw where one of the Italians had carried over a credit balance of $2,600, which he was leaving with the company on interest."[9]

Quackenbos, on the other hand, had no interest in such examples of Italian prosperity. Her mission was not to applaud the successful but to vindicate the weak against rapacious profiteers. She reported that the firm in 1907 had a gross income of $120,950, and, deducting $34,000 to tenants, made a return of $86,950, a sum indecently large (see table). The revenue was chiefly garnered from the acreage rented. Other profits came from commissary sales, with exorbitantly high prices for goods, as well as from rental and sale of mules, from a rebate of approximately 20 percent from the physician's fee, and from ginning, freightage of cotton, baling, and resale of seeds. Her conclusion was: "All this, to my mind, foretells an enormous profit for the company, and a heavy burden for the laborer to carry."[10]

Out of the total number of families (44, 73, and 92), those making above $400 constituted 54.5 percent in 1904, 48 percent in 1905, and 46.7 percent in 1906. Those making less than $400 per year constituted 45.5 percent in 1904, 52 percent in 1905, and 53.2 percent in 1906. The accumulated total per family averaged $1,627.98 for three years of work. The twenty who earned between $1,000 and $1,600 per year thus came close to the figure of $5,000 that Percy liked to use as an example of what could be, and actually was, accomplished—by a few. The totals, however, are less than the number actually on the soil. There were 125 families, not

TABLE 1   Amounts Paid to Sunnyside Colonists 1904–07:
          Three Crop Seasons

From a list prepared by Mary Grace Quackenbos of the number of tenants and the amounts paid to them in cash. In 1904–05 the total paid was $23,594.90, coming to an average of $536.24 per family. In 1905–06 the total paid was $45,242.90, at an average of $619.77; and in 1906–07 the total was $34,454.02, at an average of $374.50. It is true, though, that the amounts paid to individual tenant families varied widely; over half were making $400 or less, but the percentage of top winners and losers was about the same percentage of the whole.

AMOUNT PAID	$1,000– 1,600	$800– 999	$600– 799	$400– 599	$200– 399	$100– 199	$0– 99	NUMBER OF FAMILIES
1904–1905	3	8	4	9	12	8	0	44
1905–1906	10	11	7	7	23	6	9	73
1906–1907	7	10	8	18	20	23	6	92

Source: Mary Grace Quackenbos to Charles J. Bonaparte, August 2, 1907, and Leroy Percy to Quackenbos, August 17, 1907, Department of Justice Records, R.G. 60, 100937, National Archives, Washington, D.C.

44, present in 1904–05, and 158 families, not 92, present in the fall of 1907. This, then, is a sample, but since Percy's and Quackenbos's interpretations of the firm's books coincide, no dishonesty is apparent.

Yet, she was unfair. She had not properly counted company expenses: her calculations left out railroad and gin equipment purchases and maintenance, interest payments for money borrowed for company operations and for money lent the immigrants for transportation and living, the Corbin heirs' share in the profits, the salaries of the four Sunnyside employees, payments to the Italians for the clearing of lands, ditching, and other tasks, state and local taxes, and the prepayments to shipping and railroad companies for transportation for new immigrants. That the transportation scheme, with secret instructions on how to answer immigration authorities, was illegal, and that the exaggerated promises promoters and agents used were unethical are charges that were undoubtedly true. But if Percy was guilty, so were hundreds of other businessmen engaged in the practice from one coast to the other. Even Quackenbos admitted that the "general system" rather than individual planters vexed the tenants.[11] Indeed, Percy's reputation was not so grim that his name was imprinted on the memories of the immigrants who left the colony.

Pestilence at Sunnyside—the death of "perhaps hundreds" of colonists writes one historian—far outweighed recollections of coercion and rapacity.[12]

At the same time, Percy had good reason to boast of his renters' success. According to the figures that Quackenbos's final report and Percy's own remarks supplied, the Sunnyside Italians' total average gain from 1904 through 1906 amounted to about $732.56, or $244.19 per year for each family over and above all debts owed to the Sunnyside Company. According to the Department of Agriculture, farm laborers in Arkansas earned an average of $173.88 when supplied with board in 1909, three years later. Thus, on average, the Italians earned $70.30 a year more than the vast majority of native-born tenants.

If the Italians had hired themselves to farmers in Wisconsin, for instance, they would have made $293.04 with board. Californians were the most generous employers of all, with average farm laborers receiving $383.28. One must remember that the northern farmers were paid by the month, and some months the workers found no employment at all.[13] Also, without Sunnyside's garden plots, chickens, hogs, and fish in nearby Lake Chicot to help them through bad times, they faced unemployment as a constant threat. To be sure, city jobs permitted greater savings, but some Italians preferred rural life and retained a love of the land that was traditional in the home country.[14] The 10 percent interest charges at the Sunnyside commissary, about which Quackenbos and the colonists complained, at least provided a ready resource. The rate itself was unexceptional in the credit-poor South. While higher than those in stores less isolated, the prices of goods were well within range of what the workers could afford. John Savage, president of the Mississippi Board of Trade, Freight Bureau Department, wrote his good friend Percy, "There might have been some justification for the complaints against the plantation store, but I do not think there should have been any objection to the planter handling the cotton grown by the tenant, as he would probably get more money for it than the tenant himself." Percy did not, however, concede any point on this matter. He rejoined that the commissary gave credit when other stores refused it. With regard to the question of usury, he argued, "We are not willing to charge less than 10 per cent interest, this being ten per cent per annum. There are thousands of people in this country with good security, who cannot get money at a less rate of interest. The money we operate this place on usually cost us more than six per cent."[15] Unhappily, these were standard Delta practices. Percy was neither kinder nor crueler than his equally wealthy neighbors.

In that period of American history, unskilled rural folk with little education had few options. The one Percy offered was far from princely. Yet by no means was it the worst fate imaginable. Even Italian Ambassador Des Planches—before Quackenbos's investigation—admitted that the Delta plantations were preferable to conditions in the northern Italian provinces from whence the immigrants had come.[16] Needless to say, there was plenty of room for overcharging and unscrupulousness, especially on cotton sales at less than the going rate. "The tenants complain," reported Quackenbos, "that the Company pays them three or four cents less per pound than other planters in the vicinity and a much lower figure than that to be had at the Greenville market." Probably this was so, but in hopes of ending the controversy, Percy agreed to offer a more equitable way of handling cotton sales.[17] Having made a few other concessions, Percy argued—not unjustifiably—that the immigrant had reason for hope.[18]

Quackenbos denied such assertions but left herself open to counterattack. Progressive and northern in attitude, she used the most harrowing examples that she could find, a policy that exposed her to charges of romantic impracticality.[19] She began the inquiry with high hopes of securing indictments for peonage, which she defined as advancing money "as a means to bring over to this country contract laborers in defraud of immigration laws and regulations of July 1, 1907."[20] Yet that objective became almost secondary, in part because she at first unearthed no unambiguous cases of outright peonage to prosecute.

Frustrated and isolated, with adversaries on every side, the New York reformer turned somewhat from matters of criminal offenses. Instead, she concentrated upon plans to improve the operation in light of conditions she thought wretched if not exactly unlawful. As a result of her quickly broadening agenda, Quackenbos did not devote sufficient time to financial matters, an essential part of a legal investigation into peonage. Most important, she did not fully uncover how the less-fortunate settlers had fallen into a morass of debt and how long they had remained there. Among the 158 families, she identified 59 debtors. Yet, 35 percent of them were newcomers who had not received cash for the 1907-08 crop then being processed. Benjamin Humphreys, Greenville's congressman, reported figures similar to those that Percy produced. Quoting from Mississippi planter Alfred H. Stone's speech to the American Economic Association, Congressman Humphreys pointed out that "out of the 110 Italian squads who started to work at the beginning of the current year [1905] 44 were new arrivals." Yet sixty-five squads, or 59 percent of the total number, had no remaining debts at all after settling their accounts when the cotton was sold. They had reached "a condition of independence" in that single year.[21] The same, Humphreys could reasonably surmise, was likely to be true for the latest group of newcomers. Of those whom Quackenbos identified in her 1907 survey, fifteen, a quarter of the fifty-nine, owed less than $250. That sum was the approximate average amount their cotton would likely have been worth.[22] The result spoke well for the company's management, but certainly even more eloquently for the majority of hardworking, thrifty colonists themselves.

Lambasting Quackenbos's reform instincts as sheer sentimentality, Percy effectively challenged her grasp of the whole enterprise. The lawyer told the press that the reason she had taken up the case against O. B. Crittenden for peonage was that he had forbidden her to continue investigations at Sunnyside. "When I found that they were interfering with and demarlyzing [sic] the labor, both white and black," Percy recalled, "I notified her that their remaining upon the property would be treated as a trespass. This she resented, and this is the foundation of the peonage case. I had to choose between offending the lady or having the labor on the property completely demoralized and chose the former alternative."[23] Captain J. S. McNeilly, the editor of the Vicksburg Herald, in which Percy's statement appeared, could always be counted on to defend the Sunnyside Company. The old Confederate veteran had long been a family ally, a relationship that was cemented by the editor's marriage to Leroy's aunt, May Percy. Such kinship ties as McNeilly and Percy had always strengthened Delta opposition to outside intervention.

Mary Grace Quackenbos also made a number of tactical blunders that lost her the respect of her adversaries. Intelligent and forthright though she was, the New York attorney knew nothing of cotton operations, as Percy complained. Naively, she thought that, because of two periods set for ginning, two crops were raised annually.[24] In a letter to Charles Bonaparte, the United States attorney general, Percy argued that Quackenbos "does not come as an impartial investigator, as an unbiased judge, but, as she says, simply as an advocate of the poor." She overlooked the prosperous farmer to focus her "womanly sympathies" on the "unfortunate or sick laborer."[25]

Nor were all her suggestions really practical, as the Sunnyside managers protested. For instance, she argued that the Italians should have access to the marketplace, gins, and other needed paraphernalia of their cropping, but Sunnyside was too isolated, the costs too high to search for distant competitors, their English too imperfect, and the native whites too contemptuous of foreigners for them to negotiate without getting cheated abroad. Leaving aside his obvious antifeminist bias, Percy had grounds for vexation.

Yet, much that Quackenbos proposed was quite practical and useful, particularly with regard to health and living conditions. Unfortunately, her sympathies did outrun her authority as a federal prosecutor. She was only supposed to gather affidavits related to alien contract labor law violations. With some cause, Congressman Humphreys pointed out her unlawyerlike zeal. At the same time, no doubt to the amusement of his northern listeners, he accompanied his complaint with much rhetoric about the "high-spirited, proud, quick-tempered" men of "chivalry" who learned their veneration of women "from their cavalier ancestors."[26]

Despite the disrespect shown Quackenbos, despite the irrefutable evidence of illegality that Lewis Baker denies, despite the validity of peonage charges, Percy and other Delta planters had to have labor.[27] At risk of violating the protectionist immigration restrictions, Percy either could get the labor he needed abroad or leave the land fallow or unimproved. Blacks were moving off the land to find work in lumbering, mills, mines, and jobs in southern cities. Many left because of the rising tide of racism that James Vardaman and other demagogues were helping to arouse.[28]

Under these circumstances, to violate the alien contract labor laws was more than a casual temptation.[29] Debt peonage did exist at Sunnyside— in the form of the prepayment of transportation from Italy and the distribution of "show money" to the Italians in order to win entry through immigration barriers. Since the immigrant could seldom pay for transportation and live for a year independent of the employer's capital, the enterprising landlord had to cover the workers' expenses. If the laborers fled those debts, owners would lose their entire investment, firms would go broke, and successful renters would fail along with landlords.

The Greenville attorney owned Trail Lake, another plantation that he ran without partners near Arcola, Mississippi. The year before Quackenbos's arrival, the Italian settlers had fled, leaving behind an indebtedness totaling twenty thousand dollars. Even at Sunnyside, some thirty workers had absconded without meeting their debts, whether these obligations were exorbitant or fair. On the whole, the colonists were doing too well to sustain the charges, despite the coercion employed. Attorney General Charles J. Bonaparte put that issue aside, claiming in a letter to Percy that "the question is not whether any particular 'colonists' are, or are not, prosperous, but whether they are, or are not, free, as the law requires them to be."[30] In effect, Bonaparte was arguing that the

Thirteenth Amendment of the Constitution was applicable here. Yet, contract law then, as now, required fulfillment of all obligations agreed to. Professional athletes, actors, and others sign contracts restricting or prohibiting their freedom of choice of employment for a stipulated period. Owners of entertainment companies and athletic clubs buy the labor at terms lucrative enough to make obedience to the limitations worthwhile enough so that in a legal dispute charges that the antislavery amendment had been violated would scarcely impress a jury. To be sure, the Italians had serious grievances and were mistreated, but an attorney like Percy could argue that hardships were inevitable in farming life and the rewards, modest though they might be, made them bearable.[31]

During the crisis, Percy urged investigation by a competent and "fair-minded" federal agent.[32] Such an exposure, he thought would reveal the prosperity of the Italian settlers on his properties. Also it would serve as a model of worker and employer relations and revive the southern rural economy.[33] Indeed, if things were as bad as the northern press and Quackenbos claimed, why did the colonists "constantly demand the bringing over of their relatives and friends," a query that Percy registered with the Italian ambassador. (Percy invariably referred to them as "colonists" in the press but "tenants" in private correspondence.)[34]

From Percy's point of view, suitable alternatives seemed quite unavailable. At a low point in morale in the spring of 1907, he had written Father Galloni, the resident priest at Sunnyside, how sad that colonists should leave the land simply to work in the coal and iron mines of Birmingham, Alabama. "There is no harder work any where in the world than in a coal mine, none less suited to agriculturalists, and there is no future in it," Percy argued.[35] Given the wages that were paid to nonunion, southern miners, he had a point. Moreover, as Percy remarked to Quackenbos, "A large number of Italians who were perfectly able to go anywhere they wished" had decided to remain on the property. At least their decision showed a willingness to deal with conditions as they were.[36]

To be sure, Percy's chief object was making money. Yet he also entertained grander hopes, as Lewis Baker claims. One might call him the Henry Grady of Delta agriculture. Just as the Atlanta journalist sought to industrialize and urbanize the South, Percy aspired to bring order and wealth to the regional countryside. Indeed, throughout southern history, at least until more recent times, agrarian ventures had always taken priority over manufacturing. Time and again, Leroy Percy insisted that the Delta was a rich land with only a few drawbacks. First among the admitted disadvantages, however, was its reputation for insalubrity. Percy readily admitted that underdeveloped "semi-tropical" lands were bound to be malarial. Yet, he asserted, "There is no typhoid fever, little pneumonia, and little consumption." Up and down the Delta, pump water, he insisted, was what nearly everyone drank. For the Italians to demand artesian water was simply impractical, Percy maintained. Nonetheless, his defense should be taken with a grain of salt. His refusal to screen cabins because of the expense, for instance, was shortsighted, if only from a public-relations viewpoint. The water was polluted and should have been improved. Medical care should have been entirely reorganized.[37]

On another issue, however, he struck closer to the truth. Given the problem of labor shortage, blacks had been "forced or persuaded," as he confessed, to cultivate more land than they could efficiently manage. Percy failed to mention another factor. Some blacks took jobs under white bosses only reluctantly, even sullenly. In some Delta districts they had made up a majority to two-thirds of all farm owners, and many had recently been flourishing. Then, low cotton prices in the 1890s had forced them to surrender their lands to the mortgage holders and to look for menial work.[38] In response to the circumstances, Percy rhapsodized about the prospects for white settlement. As many as 200,000 new white workers, he hoped, would arrive to relieve the labor shortage without jeopardizing the employment of unskilled black labor. Unfortunately, for his plans, the white-skin invasion was never to reach the Delta river banks.[39]

Despite these factors, which suggest that Percy's company was no villainous operation, his aims were deeply contradictory. Percy wanted to attract free men who were supposed to accept a degree of debt servitude in return for the chance to start life over. The chief problem was the scarcely hidden presumption that the Italians deserved to be categorized as less than full citizens; perhaps they were considered better than blacks, but not truly free men. For that reason, the experiment ultimately failed—as well it should have. The problem was less Percy's, however, than the historical and economic situation itself.[40]

Rare was the company in the South that did not have serious credit problems. It took a heavy infusion of capital to meet the costs of bringing distant labor to the edge of Lake Chicot. The money borrowed for operations was not cheap. A look at the firm's few remaining papers shows how hard it was to raise the funds. Sunnyside carried a mortgage of $300,000, which, at 6 percent, would cost $18,000 a year in simple interest charges.[41] To recover the investment, the company had only two methods available: first, to make conditions sufficiently comfortable that all the renters willingly stayed, or, second, to compel them to work. The first proposition was probably impossible. Health conditions were such that expensive improvements would alleviate the problems but not cure them.

The second method, the use of coercion, was much more congenial to Percy and his friends. In order to obtain the labor in the first place, he connived with agents in Italy, New York, and New Orleans, making promises that implied a degree of respect that the settlers would soon find wanting. The use of force, traditional in the South, was the result. As Percy told Quackenbos, "We are not operating the property as an eleemosynary institution." He proposed that the company would offer an "extremely liberal and fair contract," but it would not be one "so beneficial to the tenant that it will be ruinous to us."[42] By venerable usage, those with capital had the implicit right to extract their investment from Italian peasants just as Percy's forefathers had in working their slaves.

A plantation mentality appeared in his attitude toward outsiders. He saw to it that Quackenbos was labeled a "socialist," a familiar southern canard then and later.[43] With regard to the Italians, Percy complained that while most settlers were appropriately hardworking and docile, some, regrettably, were not flourishing. The latter had more children and infirm elders than the requisite breadwinners to feed them. Others were inexperienced newcomers to the cotton fields. But beyond that, his prejudices were typical of his era and region. Thanks to their "excitable nature," Percy fumed, the Italians had the insufferable gall to "pour out their woes to any willing listener."[44] This son, grandson, and great-grandson of slaveholders was quite alarmed at such displays of protest. He was used to the pliant demeanor of long-oppressed blacks and sometimes almost regretted his decision to recruit Italians. Only "a regiment of Pinkerton Detectives" could prevent the Italians from stealing the Sunnyside cotton, he once wrote in exasperation. "They are industrious, but no more honest than the negro, and much more enterprising."[45]

In making his defense of the beleaguered company, Percy used all the resources that the southern planter had always employed. At first he tried to woo Mrs. Quackenbos with southern charm and gallantry. Briefly, it worked. She remarked, "Mr. Percy appears to me to be a man of common sense" at least compared with the "bullying Mr. Crittenden," who, she added, "is without heart."[46] Unused to federal intervention and confident of his mastery of the situation, Percy introduced her to the gracious and beautiful Camille, his wife, gave Quackenbos a letter of introduction to another planter, and allowed her access to Sunnyside. Quackenbos was soon disabused. She wrote Percy: "You have endeavored to insult me several times during the course of my investigation for the United States Government upon your Sunnyside property" and called on the managers to do likewise. "I am amazed to learn that you were the originator of such ungentlemanly behavior!" Confident that nothing serious would emerge, he then went off for several crucial weeks in the late summer of 1907 to hunt elk and fowl in Yellowstone Park. It was an expensive vacation, no doubt paid for by the sweat of his Italian as well as his black menials.[47]

In October 1907, matters took a more serious turn: a federal grand jury indicted his partner O. B. Crittenden for peonage. Crittenden and a local policeman had forced two young Italians, who were heading for the mines outside Birmingham, to leave their train and had returned them to Sunnyside at gunpoint. Percy swung his political weight with mighty effect.[48] First, he sought relief at the White House. Percy and his Sunnyside partner Rosenstock, whom Quackenbos identified as "a wealthy Hebrew," had entertained Roosevelt on a Delta bear hunt in 1902. Initially, the president was appropriately noncommittal.[49] Later, Roosevelt took a different line. When Percy visited the White House in November 1907, he found the president "courteous and polite," even "friendly." The president informed his visitor that he had personally forbade Quackenbos from investigating Sunnyside any further. In the public press, however, Percy boasted too soon of his successes against Quackenbos, to the president's embarrassment. As a result she was briefly enabled to return to Mississippi. By the end of November, however, Percy had regained the upper hand, and she was permanently removed from the case.[50]

Roosevelt, whose mother had been a Georgia aristocrat, understood the South better than Quackenbos did—and he was willing to make excuses for its peculiarities. To his friend Albert Bushnell Hart, the president later confided that "infamous outrages," not to be countenanced "in Oyster Bay or Cambridge," took place in the South with singular regularity. In that part of the world, "the habits of life, the sentiments of the people" suggested a different age and order of things. Such factors had to be taken into account "in endeavoring to enforce laws which can be enforced save by juries."[51] Loyal partisan though he was, Roosevelt had no intention of resorting to the politically disastrous policies of "Black Republican" Reconstruction of forty years before. With Quackenbos specifically in mind, the president wrote his attorney general that he preferred to rely upon "absolutely fearless, upright, and honest men" of the South rather than "northerners" for investigative work. In fact, he hoped to advance Republican fortunes in the South by appointing such men as Percy to patronage posts. Throughout the fall of 1907, Percy's name was noised about for a federal judgeship in Mississippi, a rumor that scarcely helped Quackenbos's case.[52]

Having smoothed matters over with Roosevelt, Percy sought next to reach "Charlie" Lee, the United States prosecutor in Jackson. Knowing that he could not win a long struggle with Percy, Lee tried to slow the federal investigations but could not halt the well-publicized Crittenden case.[53] As Crittenden's attorney, Percy reassured himself that the relevant Mississippi statutes were constitutional. During the brief hearing, he was gratified that his tender, even motherly "nursing" of the grand jury had paid off. The foreman and jury members were friendly, "fair," and loyal Mississippians, he was pleased to report.[54] Charles Wells Russell of the Justice Department, who had taken Quackenbos's place, was unable to secure an indictment. A lack of evidence was not the problem. Rather, the grand jury had concluded that no petit jury would convict. White Mississippians, it would seem, "Knew no Prince but a Percy"—as it once had been said of the feudal Northumbrians. Needless to say, "Crit," whom Percy called a great "sport," warmly congratulated his attorney.[55]

Closer to home, Percy adopted the cruder tactics so long employed in the Deep South that to foreswear them on grounds of principle would have amused if not horrified him and most other Delta whites. Yet, it is fair to say that by the conventions of the day, Percy's rule was almost benign. The case of Charles Pettek, Quackenbos's Italian-speaking investigator whom the Sunnyside manager had caught and arrested, illustrates the point. At Lake Village, Pettek had to stand trial for trespassing on the Sunnyside property. A neighboring justice of the peace, thoroughly drunk, convicted him and sentenced him to a hundred-dollar fine or three months on a chain gang. The agent would have had a stronger case if he had shown his credentials at the start. As it was, manager Tom Wright and friends thought that the agent planned to entice away the work force.[56]

As historian Pete Daniel has shown, many more violent atrocities than any incidents at Sunnyside took place in other parts of the Delta. Although Pettek and Quackenbos were outraged, worse might have befallen Pettek.[57] Nonetheless, the situation was bad enough. It required no direct orders from Percy to have the Italians' mail rifled at the Sunnyside and Greenville post offices to find those escapees who thought digging coal in Alabama preferable to farming in the swamp around Lake Chicot. Working under the eyes of armed horsemen was itself intimidating. Moreover, Percy did not have to reveal his hand when his old drinking friend and political ally, Colonel Thomas C. Catchings, "recovered" Mrs. Quackenbos's incriminating papers. She found them missing from her room at the Cowan Hotel in Greenville. They contained the Birmingham addresses of potential witnesses—escaped peons, Quackenbos called them. Knowledge of their new location was highly significant to the burglars for obvious reasons. Catchings had ordered the records returned, one would guess, as a greater insult and as a more ominous threat than if they had simply and mysteriously disappeared. Quackenbos suspected that Percy was behind the purloining and was misusing his local power. "Mississippi and Arkansas regard Percy as a 'hero,'" but as far as she was concerned, "he is *not on the square*" and is "committing crimes against the Government cleverly hid."[58]

Percy himself seldom if ever authorized the use of unlawful means, but never disowned ill conduct when others did so. Clearly, immigration agent John Gruenberg felt most uncomfortable during an interview with the Greenville leader in November 1907. Percy, he reported, had been friendly enough until mention was made of using Mrs. Quackenbos's interpreter for his Sunnyside inspection. Irritably, Percy rejoined that such an arrangement would reduplicate the alleged misstatements of Quackenbos and would lower plantation morale still further. Gruenberg reported home that he could not vouch for the interpreter's life (and implicitly his own) if the pair ventured on the property.[59] Having won his point, Percy then expressed approval of the seated bureaucrat from New York. How much easier it was, he undoubtedly mused, to deal with a *man* instead of a woman who had boldly appropriated Percy's sense of chivalry as a buckler against his wrath. Gruenberg was aware that he had no such shield.[60] But, the agent reported, he knew "enough to avoid collisions with the prevailing adverse sentiments."[61] As a result, in contrast to Quackenbos's persistence, he did nothing at all.[62]

By these customary strategies, Percy won a number of victories. He sailed easily past the clumsy grasp of federal law, sprung his partner from its traps, emerged as heroic champion of Delta interests, and saved the Sunnyside Company from the reforms that Quackenbos hoped to achieve.[63] But these triumphs were only Pyrrhic. Ultimately, he had run into an insurmountable problem. Armed with the Quackenbos report, the Italian government, through its local consuls, prohibited labor agents from recruiting workers for the southern states and even the tenants themselves from bringing over friends and kindred.

Percy suspected that other southern states, especially Texas, sought to steal Italian labor away. The Italian government "has taken a decidedly hostile stand against Mississippi, and the Delta in particular," Percy wrote early in the controversy. "I don't know that the officials have been improperly influenced, but they are bending all their energy to diverting this immigration to Texas, and throwing around the immigration to Mississippi all the restrictions in their power, and they are virtually able to stop this immigration to this section."[64] Mary Grace Quackenbos's full report, which her own department failed to retain in its files, did have an impact upon the Italian authorities who requested it from the State Department. Making rigid new rules regarding immigration documents, they successfully halted illegal contract labor arrangements in both America and Italy, thanks to the evidence and testimony that Percy's adversary had gathered. In the meantime, in Italy and the American ports, word spread that on Sunnyside and the nearly one hundred other plantations up and down the Mississippi, peasants were being fashioned into slaves. Law enforcement and rumors, not criminal prosecution, ruined Percy's operation.[65]

By the close of 1907, the Greenville attorney must have felt traduced. He had to fight two governments, his own and Italy's, the Italian colonists, who proved so disappointingly susceptible to outside agitators (as Percy regarded the authorities), and, finally, even many fellow Mississippians.[66] The latter had no use for "dagoes" or their papist religion. Against such forces, Percy could only repel attacks. He could not win the campaign. Finally, in 1913, he sold out his share of the Sunnyside properties. The plantation vainly struggled on, disastrously adopting the sharecrop system. The Italians vanished. No one took their place.

The failure of the Delta experiment was most unfortunate. The South needed the kind of pluralism that was fast making the North, its former enemy, so rich and powerful. But, in the last analysis, Percy's defeat cannot be blamed on his enemies at home and abroad. Nor can his own role in the business be entirely laid at the environmental doorstep. Instead, a failure to rise above his time and place was evident. He could not adjust to the new order that he sought to build. Not once, if the records do not mislead, did he ever talk to the Italians themselves. In contrast, John

Gracie of New Gascony, another very wealthy Arkansas landlord, had cooperated with Quackenbos and undertook most of the reforms she suggested—without going out of business. He even learned some Italian.[67]

Percy was too arrogant to bend, particularly to a demanding woman. Nor did he adopt his customarily wide sense of public responsibility to include the needs of the Italian newcomers. For instance, among his many contributions to Greenville culture and civic life was his support for the city's school system that Superintendent E. E. Bass had molded into the best in the state.[68] In contrast, the Sunnyside owner provided nothing for the children of the Italian workers. As Ambassador des Planches observed, "Un *drawback* grave di Sunny Side è la mancanza d'una scuola"—the lack of school for the two hundred or more children there.[69] A little more paternalism, a little more chivalry, if you will, might have helped. Instead, Sunnyside residents felt as if they were cogs in a great, dull, agrarian machine—"una machina umana di produzione."[70]

Finally, Percy never offered the land for sale to successful tenants. The blame, however, did not rest at his doorstep. In fact, the opposite was true. Percy once appealed to George Edgell, the Sunnyside Company's president in New York, "I know the experiment of selling to them was tried once before unsuccessfully, but at that time they were without means, and, as I understand it, the land was sold on credit at quite a high price and cotton was very low." Percy urged that about forty or more Italian land purchasers could "become a nucleus, which could be widely advertised, and probably serve to draw others to the property, and they would also serve as an anchorage to the tenants on the places unable to purchase, holding out to them the possibility of acquiring land in the future by staying upon the property and successfully cultivating it."[71] Apparently such a utopian idea did not appeal to the Wall Street businessman.[72] In making his case, Percy failed to point out that, with several million Delta acres still wilderness, the firm could have applied some of its profits to their purchase and development.

Indeed, the experience of the Sunnyside Italians was scarcely different from that of their compatriots who had emigrated to the coffee plantations west of São Paulo, Brazil, whom Thomas Holloway has studied. Both groups made their first priority the acquisition of land, were deeply frustrated when that opportunity was denied them, and seized every chance available to reach that dream. Immigrating chiefly in large family groups, these newcomers to the American South and to Brazil held their expenses to a minimum and sought to remove themselves from indebtedness as speedily as possible. Both found conditions harsh, death a frequent guest, and labor too often unrewarded. Yet, in both hemispheres, the colonists did more than endure. In many cases, they thrived, but with little reason to thank such landlords as Leroy Percy, except that these landlords provided such fertile soils.[73]

Like his Brazilian counterparts, also sons of slaveholders, Percy relied on old methods: surveillance by bull-necked police, crooked magistrates, and hard-minded plantation managers; political manipulation; and sometimes veiled personal threats.[74] Pride and rigidity, as well as a typically southern defensiveness that precluded any admission of blunder or wrongdoing, governed his conduct. In the long run that policy did not pay. In a passage in *The Last Gentleman*, Walker Percy well described his great uncle. "The great grandfather," the novelist wrote, "knew what was what and said so and acted accordingly and did not care what anyone thought"—most especially, he might have added, a Lady Bountiful from New York.[75] More than anything else, the planter mentality, along with the lack of full vision that went with it, was Leroy Percy's undoing at Sunnyside.

Italian Migration: From Sunnyside to the World

George E. Pozzetta

> *Emigration from Italy belongs among the extraordinary movements of mankind. In its chief lineaments it has no like. Through the number of men it has involved and the course it has pursued, through its long continuance on a great scale and its role in other lands, it stands alone.*
>
> ROBERT FOERSTER,
> *The Italian Emigration of Our Times* (1919)

The saga of Italian migration to the United States has been told so often in the context of one-way movement from the Italian peninsula to America and in terms of dense metropolitan settlement that it is often difficult to conceive of it differently. Images of the urban "Little Italy," with its colorful religious processions, black-clad peasant women bargaining in the streets, Italian import stores, and long rows of uniform tenement buildings still dominate the national understanding of this great folk movement. Yet, the Italian presence in America was far more complex than these conceptions imply, a reality to which the Sunnyside experience speaks.[1] By placing the Arkansas experiment historiographically within the major contours of Italian migration, this movement is illuminated in relation to a wider pattern of emigration and settlement.

Those individuals who came to Sunnyside were part of a truly massive migration that reached global dimensions. Scholars have estimated that some twenty-six million Italians left their homeland for various destinations around the world between 1876 and 1976.[2] Additional hundreds of thousands of migrants left prior to 1876, when the Italian government began keeping official records on migration. Although it is impossible to determine exactly how many individuals were involved in this earlier exodus, it is clear that collectively this was one of the great migrations of modern times, a movement that had significant consequences for the people involved, the Italian nation, and the countries of destination.

As impressive as the size of the Italian migration was, it is worth noting that perhaps no people (with the possible exception of the Chinese) have ever engaged in such diverse migration experiences or had its presence felt in such widely separated locales. Italian street musicians plied their trade in New York City, London, Paris, and Moscow; plasterers and masons worked all through Germany, France, and the Balkans; Sicilian fishermen ventured as far as Australia and as close as the Adriatic; construction workers labored on railroads in Siberia, South Africa, and the Canadian Northwest; and agricultural colonies of Italians dotted the landscape of various nations around the world, including Brazil, Peru, Cuba, Tunisia, and Rumania. In the age before Concord jets, Italians became the world's premier globetrotters, and they did so by constructing a network of intricate migration traditions that established, put into motion, and sustained complex movements worldwide.[3]

Thus, it is essential to view settlement in such places as Sunnyside in the context of international migrations that permitted people to probe for economic opportunity from a wide range of possibilities.[4] In many instances, the choice of destinations flowed from the institutions and individuals involved in the actual process of migration itself. A number of scholars have attempted to explore the mechanisms that brought information to remote villages, energized individuals to move, provided for transportation, and channeled people along human corridors of support and advice.

The critical importance of chain migrations, including "chain occupations," in the movement and concentration of Italians in certain locales has been studied in a number of articles by John and Leatrice MacDonald.[5] Such chains linked people from specific villages and kin groups to precise new-world locations like Sunnyside. They determined that the movement of Italians was not random or carried out in the delirium of an "American fever," but was purposeful, directed, and aimed at accomplishing a variety of family goals. The workings of migration chains also meant that migration was very particularistic, striking certain settlements and towns with varying degrees of intensity. These insights have been verified in many other studies of different kinds of migratory movement.[6]

Robert Harney has subjected the network of agents, shippers, money handlers, and other go-betweens involved in the "commerce of migration" to analysis. His work has shown how important these middlemen were in mobilizing and distributing migrants in their travels. Although documenting evidence of abuse and exploitation in their dealings, Harney has gone beyond the emotional and often pejorative reporting on these brokers to illuminate the functional roles they performed in the migration process.[7] John Zucchi has explored the significance of particularly exotic forms of Italian migration, such as the movement of street musicians, chimney cleaners, plaster statuary salesmen, and skilled masons in the shaping of settlement patterns.[8] These types of tradesmen often filled the role of "pioneers" in the migration scenario and were significant in locating new opportunity environments to which greater numbers of kinsmen and fellow villagers could be attracted later.

In exploring the workings of migration, no institution has been more important in recruiting and distributing individuals than the *padrone,* or labor agent. Indeed, the origins of the Sunnyside Italian presence can be traced to the presence of a "storekeeper-turned-labor agent" who relied on a network of subagents in Italy to recruit Italian families. The middleman in this case, a person named Umberto Pierini, collected commissions from immigrants in return for arranging passage and other transactions.[9] Thousands of agents such as Pierini operated in both the United States and Italy and were critical intermediaries in the mass movement of people between the two nations.

These individuals also served as lightning rods in generating intensely negative responses from the larger American society toward Italian immigrants. Because the padrone system was often the subject of sensationalized reporting in the nation's newspapers and magazines, which stressed the degree of manipulation and abuse present, many Americans came to regard Italians as undesirable, "coolie labor." Labor unions in particular portrayed Italian laborers as herdlike drones who constituted a reservoir of pliant strikebreakers that employers used to undercut wage levels and destroy union bargaining power. Such "degraded" labor, in their view, should be restricted from entering the country.

Both Humbert Nelli and Luciano J. Iorizzo have studied the workings of the padrone system, revealing how it was able to funnel millions of unskilled laborers into various labor markets.[10] Both authors have outlined in detail the fraudulent aspects of padrone dealings with immigrant workers, revealing how immigrants were often bilked out of finding fees and sent into conditions of labor peonage. Readers should not overlook Robert Harney's insightful writings on the padrone that attempt to place this institution into the wider pattern of immigrant strategies and development. Harney has stood the traditional argument on its head by showing how, despite exploitation, the padrone system served the needs of the migrants. He concluded that padronism only ended when migrants no longer required the services it provided, and he has urged scholars to see the padrone system as "part of the commerce of migration, not as a form of ethnic crime."[11]

Just as the Sunnyside experiment mirrored Italian migration elsewhere in its padrone beginnings, so too did it connect with broader patterns in its ability to involve the governments of the United States and Italy at the highest levels. The interest shown by the Italian ambassador and other officials in conditions at Sunnyside was not unprecedented. Although the Italian government did not intervene frequently to protect the rights of immigrants, authorities in Rome occasionally took active roles. Interestingly, those events that resulted in diplomatic exchanges often involved Italian immigrants who settled in the American South, and these incidents typically took place during roughly the same time period as the movement to Arkansas. For example, one of America's bloodiest mass lynchings, the 1891 "New Orleans incident," in which a mob killed eleven Italians who had been accused of complicity in the murder of the city's police chief, was not far removed in either time or place from the founding of Sunnyside. This event has attracted much scholarly attention, in part because diplomatic relations between America and Italy were almost ruptured over Italian demands for reparations over the killings. The literature on the incident is conveniently summarized, with a provocative reinterpretation, by Richard Gambino.[12]

John V. Baiamonte, Jr., meanwhile, has provided an interesting, though polemical, discussion of a Louisiana hanging of six Italians accused of involvement in a murder.[13] Additionally, a series of peonage cases involving work camps in West Virginia and North Carolina captured national headlines and prompted diplomatic exchanges in the early years of the century. In each case, workers had been tricked by labor agents and lured to distant job sites where they all suffered severe treatment. Bosses whipped Italians who refused to work, denied them mail service, compelled them to spend their entire salaries on food and supplies, and subjected them to armed patrols.[14] In the end, the Italian government issued warnings to immigrants against settling in the South.

If scholars have explored sensational incidents of lynching and labor abuse, they have largely ignored the peaceful settlement of Italians in rural and agricultural settings. This is true despite the fact that agricultural colonies such as Sunnyside were not uncommon in the United States and elsewhere.[15] The long-lived colony at Tontitown, Arkansas, an offshoot of Sunnyside, and the immensely successful venture of the Italian Swiss colony in the California wine country have been studied, but these cases are exceptional.[16] No doubt one reason why this dimension of the Italian migration experience has been overlooked is that so few of these experiments ultimately succeeded. Overall, industrial employment and even unskilled construction work offered promise of greater immediate reward, and most migrants gravitated to these economic sectors. One estimate made in 1909 claimed that Italian laborers could save 20 percent more during eight months of work on construction projects than they could during twelve months of farm labor.[17]

Readers interested in exploring Italian settlement in rural and small town America can turn to a useful volume edited by Rudolph J. Vecoli. This work contains essays exploring various farming and plantation communities, mining towns, and railroad work camps, including a contribution on Sunnyside by Ernesto Milani that covers some new aspects of this settlement.[18] Vecoli has published a lengthy essay dealing with Italian settlement in Minnesota, which includes discussion of rural and farming contexts.[19] Valentine Belfiglio has explored the Italian experience in Texas thoroughly, a major component of which involved agricultural settlement.[20]

J. V. Scarpaci has investigated the seasonal migration patterns that annually brought thousands of Sicilians to the sugar fields of Louisiana for the *zuccarata*, the cane harvest. She has extended earlier treatments of this movement to show how temporary field hands moved toward permanent settlement in New Orleans and scattered rural locations, and how they adapted to local patterns of race relations, sharecropping, tenancy, and land ownership.[21] Andrew Rolle's work has taken as its subject those Italians who settled west of the Mississippi. His broad-ranging examination includes coverage of Italians in mining, cattle ranching, agriculture, and fishing. With an argument that stresses the power of environment, Rolle has contended that these immigrants were "upraised" rather than uprooted, thereby undergoing a profoundly different experience from those of the Northeast.[22]

The fact that the overwhelming mass of Italian immigrants seldom chose rural, agricultural settlement options profoundly troubled many contemporaries. Numerous critics observed that the type of social evils associated with densely packed "Little Italies" (congestion, high rates of crime and disease, family disorganization, juvenile delinquency, and the perpetuation of foreign ways) could be alleviated by "colonizing" Italians on the land. Indeed, reformers often held up experiments such as Sunnyside and Tontitown as examples for future emulation, and a number of organizations came into being that were designed to foster the breakup of urban concentrations.[23] The Labor Information Office in New York, for example, attempted to meet immigrants on arrival and explain to them the range of farming opportunities available in an attempt to divert them from the city. Begun in 1906, the office closed its doors within three years. Similar efforts to sponsor distribution to rural areas failed miserably, usually in a year or less, since they did not fit the goals of the migrants themselves.[24]

The wages available in urban areas proved far too attractive to Italian immigrants, since the great majority intended to stay in America only a short time. Most people hoped to find quick employment, save ruthlessly, and return to the homeland with a substantial nest egg. These strategies profoundly shaped the adaptation process they underwent while in America. The usual pattern witnessed male laborers venturing outward from home villages in Italy for seasonal "campaigns" to earn money and return home. Italians manifested some of the highest repatriation rates of all immigrants. The reconstitution of families in new-world settlements typically came about only after a period of male sojourning, and multiple trips back and forth were common. The ability of Sunnyside to attract entire families intent on permanent settlement was, in fact, rather unusual.

Betty Boyd Caroli has investigated the phenomenon of return migration in a brief volume that is built around a series of interviews with returnees. Caroli has concluded that the repatriates had little political or social impact on Italy, but they did make an economic difference by infusing large amounts of capital into the Italian economy.[25] The exact nature of the difference repatriates made is investigated by Francesco Cerase, who found that returnees invested heavily in small plots of land in their hometowns, decisions that actually reinforced the traditional landholding and agricultural practices.[26] The negative impact of land purchases is also documented by Dino Cinel, but his work has suggested that repatriates came back with "new ideas" and an enhanced sense of personal worth.[27]

The presence of Father Pietro Bandini at Sunnyside and his promotion of the successful move to Tontitown raises questions about the nature of the Italian encounter with the Catholic Church in America. Most scholarly work on this broad topic has focused on the activities of priests and parishes in urban settlements.[28] Here, the record is mixed. Some investigators have stressed the positive, integrative role played by the Church in easing the difficult transitions of newcomers.[29] Others have been less sanguine, citing the divisive influences of the intense regional differences existing among Italians in disturbing parish harmony.[30] Still others have pointed to the lengthy record of bruising confrontations between Irish and Italian Catholics flowing from differences in Irish and Italian religious values.[31] Not to be overlooked in this debate are the activities of two dynamic leaders who founded religious orders that specifically ministered to the needs of Italian immigrants, Mother Frances Cabrini and Bishop John Baptist Scalabrini.[32]

Although studies of individual parishes and cities reveal disparate outcomes, much of the scholarly literature has pictured the Catholic Church as an indispensable institution in assisting immigrants. These views should be balanced by Rudolph Vecoli's arguments, which have questioned the importance of the Catholic Church in the lives of Italian immigrants. Vecoli has concluded that typical southern Italian peasants—especially the men—had little reverence for the Church as an institution and that the Church failed to play a major role in integrating immigrants into the wider society.[33] One case study documenting such conditions covers the experience of Italians in a southern location—Tampa, Florida. Because of its strong leftist elements, this community resisted all efforts by churches (Catholic and Protestant) to play a meaningful role in community life.[34]

Finally, what can be said of the migration experience from the perspective of the individuals involved? Since so many of the participants in these folk movements were illiterate, scholars once thought that they left no written records documenting their motivations and aspirations. Such notions can no longer be accepted. Clues to the more personal dimensions of the experience can be found, for example, in the immigrant press.[35] Notices of the birth, maturation, and death of hundreds of settlements such as Sunnyside can often be followed in the pages of immigrant newspapers. Many papers contained special pages devoted to letters, notices, and reports sent in from smaller settlements located around the nation. Often called *piccola posta* (little post office), these columns supply insight into the thoughts and emotions of immigrants as they confronted the realities of the New World. Another source is the limited number of immigrant memoirs that have been saved and published.[36] Perhaps the most poignant of these is *Rosa: The Life of an Immigrant Woman*, which describes the incredible heroism of a simple peasant woman from northern Italy who migrated with her husband to mining towns in the West and to urban centers in the Midwest.[37] Some collections of oral history have also managed to capture the intensely personal aspects of the immigrant experience.[38] Finally, the ability of both autobiography and the novel to reach the inner dimensions of immigrant lives is immense and probably underappreciated.[39]

Efforts to reach these aspects of the immigrant world provide insight into the ways in which migrants themselves were actors in the great drama that their collective decisions to move truly constituted. At Sunnyside and elsewhere, they were not faceless masses responding passively to larger forces and authorities, but rather thinking human beings who attempted to fashion real-life strategies to cope with the constricted opportunities they confronted at home and abroad. In this sense, those hardy souls who came to Arkansas joined company with millions of other newcomers who ultimately settled in America, changing themselves in the process, but also altering the nation to which they ventured.

Appendix 1: Commentary

Pete Daniel

The articles in this volume by Randolph Boehm, Ernesto Milani, and Bertram Wyatt-Brown originated as a program on peonage and the Sunnyside Plantation at the annual meeting of the Organization of American Historians held in Washington, D.C., in March 1990. With his essay on the century-long evolution of Sunnyside, Willard B. Gatewood has vastly expanded the context of the brief but explosive incidents in the first decade of the twentieth century. George E. Pozzetta, meanwhile, in his essay on Italian immigration worldwide, has provided yet another way to understand and interpret the events at Sunnyside by placing the Sunnyside Plantation in the framework of world economic transformation.

Without the contributions of Pozzetta and Gatewood, it might seem that the Sunnyside experiment with Italian labor was an isolated episode and that the plantation was a wilderness when Italians began settling there in the 1890s. In fact, by then it had accumulated a long and distinctive history that had attracted such famous southern names as Hampton and Calhoun. One of Arkansas's most powerful antebellum planters, Elisha Worthington, has also reappeared, along with his heirs— the children he fathered with a slave. The fact that certain aspects of history get buried—Worthington's love story, Corbin's convict experiment, and the Sunnyside Italian peonage episode—raises questions about selected historical memory and what purposes it serves. If we followed Leroy Percy's admonition, we might look only to success stories and lose sight of the tension that so often fuels past events. Sunnyside intersected with almost every major aspect of southern history: slavery and the plantation system, the Civil War, railroad development, convict labor, immigrant labor, New York investors, peonage, and, finally, a New Deal agency designed to aid displaced farmers. If Gatewood supplies rich context and demonstrates how history can be packed into one small locality, Pozzetta illustrates how that same small locality reflected events in a massive migration of global dimensions.

The other essays revolve around the Italian settlement on Sunnyside and the charges of peonage that set up the confrontation between Leroy Percy and Mary Grace Quackenbos. Returning to the subject of peonage after nearly twenty years has convinced me once again of its importance in the study of labor control in the South. I am surprised, given the history of peonage in this country beginning with the Mexican War and continuing through a Supreme Court case in 1988, that historians have neglected the subject. Peonage thrived in the midst of free labor, using debt as the lever of coercion. During the Progressive Era, federal courts began defining peonage, and, as the press spread sensational exposés, it became a national scandal. Two years before Mary Grace Quackenbos offended Mississippians with her investigation of Sunnyside in 1907, the United States Supreme Court ruled in *Clyatt v. U.S.* that the peonage statute of 1867 was constitutional. In 1908 a three-year effort began that led to the *Bailey v. Alabama* ruling (1911), which outlawed Alabama's "false pretenses" peonage law, and in 1914, in *United States v. Reynolds*, the court disallowed the criminal surety laws of Alabama. The first fifteen years of the century, then, were active with investigations, court cases, and massive evidence of peonage. Most of the legal activity focused on cases involving African-American farmers.

Largely due to Mary Grace Quackenbos and Charles Wells Russell, immigrants became the focus of peonage investigations between 1906 and 1908. The two tracked labor agents and immigrants from New York to Florida, Alabama, North Carolina, and Mississippi. They found convincing evidence of peonage, horrible working conditions, and violence. They concentrated upon F. J. O'Hara, a Florida turpentine producer; Francisco Sabbia and Edward J. Triay, New York labor agents who had enlisted immigrants to work on the Florida East Coast railroad; and William S. Harlan, manager of the Jackson Lumber Company in Lockhart, Alabama. Only in the Harlan case did they secure convictions. In the context of this legal activity and even the other immigrant cases, Sunnyside was but a footnote, producing only one grand jury hearing that did not result in a true bill.

Unlike the majority of peonage cases that involved immigrant construction workers, turpentine still workers, or loggers, Sunnyside involved farmers. Agricultural operations in the South resonated with cases that involved African Americans more than immigrants. In cases of both kinds, most employers accused of peonage were community leaders, many who ran large plantations and who used their position to elicit sympathy from the community and support from the press. Sunnyside is different primarily in that it rested on a long-term Italian labor experiment and because it involved Leroy Percy.

Important new evidence relating to this case has emerged since I first went through the peonage files in the National Archives. Randolph Boehm has recently turned up Mary Grace Quackenbos's Sunnyside report, buried, as Leroy Percy intended, in State Department files; and several years ago archivist Cindy Fox called to my attention the previously unavailable report of Michele Berardinelli, a special Justice Department agent who investigated Sunnyside. These documents not only show that Italian immigrants were held as peons but also expose the social and economic conditions in the Delta. Berardinelli even posed as an itinerant photographer and secretly took dozens of photographs showing the living and working conditions of the immigrants. Ernesto Milani has used Italian diplomatic correspondence to trace the development of the colony and document the Italian government's role. Finally, Bertram Wyatt-Brown has explored the Percy papers, recently liberated from the family cotton compress in Greenville, Mississippi, to determine Leroy Percy's role in this episode.

Milani explains that Sunnyside began this chapter of its history as a semi-Utopian experiment that would encourage Italian farmers to become southern yeomen and allow Austin Corbin to retire his mortgage. Despite repeated complaints about everything from contracts to drinking water, the Italian government took little action to protect the immigrants. Evidently Rome ignored reports of the colony's problems just as Washington did later when Sunnyside went under the management of O. B. Crittenden and Company. Neither government seemed disposed to take complaints seriously. By the turn of the century, the original Italian farmers melted away, searching for work in cities or settlement in Tontitown and in Missouri.

When Percy, Crittenden, and Morris Rosenstock took control, they invented a brilliant but illegal scheme to lure Italians to the colony and keep them there. Sunnyside thus became another chapter of the immigrant experience in the land where it never snows. It did not take long for the Italian embassy to hear tales of abuse.

What is important in the overall story of immigrant peonage is that the Italian government did anything at all, especially if one accepts as correct Michele Berardinelli's characterization of the consular agent in New Orleans as feeble-minded and incompetent. Professor Milani could strengthen his essay by probing deeper into the effectiveness of such officials and by comparing them to consular agents in Florida. From the evidence that I have seen in the peonage files, no government called for an investigation of Polish, German, or Russian immigrants during this time, although Berardinelli mentions Austrian intervention. So even the rather belated and inept role of the Italian government was important, and it did have some impact.

The Italian government's request for an investigation set up the confrontation between Leroy Percy and Mary Grace Quackenbos described in the essays of Bertram Wyatt-Brown and Randolph Boehm. We see here the interaction of two dynamic personalities, the intersection of two traditions. Percy versus Quackenbos. Percy embodies the southern-planter past, but he dreams of Taylorizing the Delta labor system by successfully using Italian farmers. The undermining of traditional share-cropping arrangements negotiated over the years by blacks is implicit in Percy's new corporate model. African Americans would never have submitted to such controls over their crops and seed money. Does Percy see himself as a patrician Progressive, perhaps even as a corporate (as opposed to social) reformer? In his mind he personifies the enlightened (up-to-date) New South planter, but, we learn, he has serious flaws.

Mary Grace Quackenbos personifies the social-reform spirit of the North, and she bravely marches south to stamp out peonage among immigrants and to bring the guilty to justice. She is as dedicated and as self-assured as Percy. It would be difficult for a novelist to invent two more likely antagonists. Boehm has rescued this remarkable woman from obscurity and convincingly presents her side of the story. When these characters collide at Sunnyside, it is Leroy Percy who is stunned and then panicked into massing his considerable power to discredit Quackenbos, to destroy her report, and to evade peonage charges. His aggressive assault on her raises questions not only about his motivation—to protect his investment at Sunnyside—but also about his understanding of his role as a southern gentleman. Using "persuasion and coercion"—his words— suggests that free labor and incentives were not working.

Percy used "voodoo economics" in claiming that some Italian families made up to fifteen thousand dollars over eight years, and in insisting that they were better off than blacks, indeed, better off than they were in Italy. Quackenbos was not moved. She offended Percy almost as much by her refusal to ride in his carriage, use his interpreter, and see Sunnyside through his eyes as she did by her report. Percy comprehended too late that she could not be courted in the usual southern gentlemanly way or be distracted from her mission. She would not dwell on the success stories. Instead, she diligently attempted to understand the cotton culture, and her report is filled with useful economic and social information, as is the report of Michele Berardinelli. Percy's argument to President Theodore Roosevelt that Quackenbos did not understand the cotton culture masked his fear that she understood it all too well. Had she not found things that deeply worried Percy, he could have let the matter pass. She discovered that Sunnyside prevented the movement of free labor, that Italians had become peons. Quackenbos drove him to take rash actions that contradicted his claim to chivalry and honor. He sought to dishonor a woman who was only doing her job in a professional and highly competent manner. He had, in other words, met his match in Mary Grace Quackenbos.

Percy's dismissal of Quackenbos's class consciousness as "philanthropic humanitarianism" and Roosevelt's comment on her lack of judgment as "hysterical and sentimental" serve to eclipse the serious problem of federal peonage violations. But while the president expressed puzzlement over southern "outrages" from a "different age" that were immune from law enforcement, his Justice Department was at the very time investigating and prosecuting dozens of peonage cases in the South. Such a self-serving statement from the president suggests his eagerness to accept Percy's fabricated tale of harassment at the hands of a woman.

It is unclear in Wyatt-Brown's essay whether Percy is a slave to the past or simply a Progressive Era businessman with a southern rural taste for exploitation. Does understanding the complexity of Percy's past make him simply part of the landscape and lift the historical burdens that he should bear? Profits are a thin shield to protect honor, nor do they go far in explaining the transition from slave to free labor, Mississippi style. What should a historian make of a man, and a Percy at that, who violates federal law in the operation of his plantation, condemns the woman who discovers it, defends his partner, and, of course, himself before a grand jury, winks when a friend turns up with Quackenbos's legal papers stolen from her hotel room, uses his power to discredit her, peddles his influence in the White House to have her removed from the case, prompts the Mississippi congressional delegation to further attack her (and defend him), and manipulates the press to appear as a great defender of his state and its heritage against socialist outsiders? Is this what being a Percy was all about at this moment in Mississippi history?

Wyatt-Brown dwells upon Percy's interpretation of the Italian farmers' profits, but he ignores Percy's profits and the means, some of them illegal, that he was willing to use to make them. Wyatt-Brown is difficult to read on this question, for he appears both to defend and to chastise Percy; he is an expert at discovering silver linings.

Quackenbos, in Boehm's estimation, conducts herself as a professional investigator, one who will not be sidetracked. She interviewed 70 of the 157 families on the plantation. She found a host of problems, as Boehm points out, but compared to the Florida-Alabama turpentine firms and the labor camps on the Florida Keys, Sunnyside was not particularly violent or oppressive. To her credit, she negotiated thirteen changes in the tenant contract, a remarkable achievement that certainly puts her investigation in a positive light.

It may well be true that Percy needed labor; all Delta planters cried for labor. The idea behind free labor, as I understand it, is that it pays wages. Percy charged the immigrants for transportation, sold them mules at inflated prices, used the payoff system, marketed their cotton at some 40 percent per pound less than they would have received in Greenville, kept their seed money, forced them to trade at his commissary (at 10 percent flat interest), screened their mail, prevented them from leaving until they paid their debts, and when they did try to leave, had them arrested and returned. Black sharecroppers had fought since the Civil War against such treatment, but Percy was quite willing to abolish tradition and custom in his role as a New South manager. These are, I suggest, shady practices—except for peonage. That violated federal law. Wyatt-Brown insists several times that the Italian farmers had hope. Given Quackenbos's and Berardinelli's investigations and the number of Italians who fled Sunnyside, where is the silver lining in this instance?

Percy agreed to some changes, but when the charge of peonage emerged, he claimed that Quackenbos had exceeded her authority. A peonage charge is a serious matter. Peonage is a federal offense, punishable by a fine (one thousand to five thousand dollars) and imprisonment of one to five years. The grand jury did not find a true bill against O. B. Crittenden, although the evidence offered in these essays shows a clear violation of the law. The class system of Mississippi might protect the wealthy Crittenden, but as Boehm points out, the riding bosses and corrupt law enforcement officials might not have fared so well. No wonder Percy got busy stifling Quackenbos with every weapon he commanded.

I am perplexed by the statement that if indicted, Percy "would have been acquitted, even in a jurisdiction less sympathetic than Mississippi" when Wyatt-Brown argues that Italians were "doing too well to sustain charges, despite the coercion employed." Peonage is not equivalent to doing well; it is simply keeping a person working because of debt. The law was not aimed at those doing well; it seeks out those who are forced to work because of debt. Occasionally southern juries did convict community leaders who had considerable influence. The question is not whether a jury would or would not convict Percy but why he felt it imperative to marshal such power to make sure that it did not go that far. Such actions suggest a grave concern that further investigations could mean his undoing. As to the labor shortage, is it fair to dismiss Percy's role because "the problem was less Percy's, however, than the historical and economic situation itself"? It seems to me that Sunnyside and its problems were Percy's; it was his experiment; it was his policy to have those owing money returned to the plantation and not let them leave. Where does historical determinism end and Percy's responsibility begin? Is his problem the inability to escape the chains of the Old South or an incomplete understanding of the New?

Percy did not end his defense with his visit to President Roosevelt to rid himself of Quackenbos. He later intimidated immigration agent John Gruenberg and his interpreter, sidetracked their Sunnyside investigation, and prompted the Mississippi congressional delegation to attack Quackenbos. By the time the Immigration Commission report appeared in 1911, Percy had taken a United States Senate seat, and he signed the majority report. Of the forty-two volumes, only seven pages related to peonage. Despite hundreds of complaints and investigations, dozens of federal cases, reams of publicity, and even a few convictions, the report whitewashed peonage. Today we might call this a very successful coverup. What is one to conclude about this southern knight, who in Wyatt-Brown's words, failed "to rise above his time and place"? There is a complexity in Percy, a darker side, that bears further analysis.

Leroy Percy continued his search for labor control in the Delta. Twenty years after the Sunnyside episode, during the 1927 Mississippi flood, he gambled on the fate of thousands of black workers huddled on the levee. His arguments sent away the boats that would transfer these shivering, wet, and hungry farm families to Red Cross camps on high ground at Vicksburg. As his son William A. Percy explained in *Lanterns on the Levee*, "He knew that the dispersal of our labor was a longer evil to the Delta than a flood" (258). Again, allegations of peonage emerged, since in some cases workers were held in camps because they owed debts. Again, no charges were brought.

These essays rescue a major confrontation from obscurity, and while the Sunnyside Italian settlement may be a footnote in the history of peonage, it epitomizes the collision of cultures, not only that of Quackenbos and Percy but that of Italians and southerners.

These essays all raise questions about how this episode fits into the flow of southern rural history. They demonstrate that forty years after the end of slavery a completely free labor system had not been implemented in the South. But why would peonage emerge in the Mississippi Delta, the gem of capitalist economic development in the rural South? If hope, opportunity, and profits were so great for Italians, why would planters resort to peonage, and did they use the same coercion to control black workers? Indeed, was the possibility of an investigation of black laborers a contributing reason for Percy to move so forcefully to halt the Italian investigation?

In the end, planters had to settle for their traditional laborers. Through diplomatic channels, Italians were discouraged from moving to the rural South. Leroy Percy may have won the battle with Mary Grace Quackenbos, but he lost the war. The Italians left, he sold off his share of Sunnyside, and Italian immigration to the South stopped. Percy's dream of a pliant nonblack work force faded. Mary Grace Quackenbos retreated to the North, and for years championed the cause of the poor and unfortunate. It was people like Quackenbos who would ally themselves with the poor and press the South a half-century later to challenge the system of segregation and discrimination that the Percys had created and for so long defended.

Appendix 2: Documents

1. 1895 Contract with Corbin.

"United States of America, State of Arkansas, Chicot county,
Sunny Side Company, organized April 19, 1897 [sic],* in the State
of Connecticut, Hon. Austin Corbin, of New York, president. Legal
representative in Italy, Don Emanuele, principe di Suasa Ruspoli,
this day of Monday, Sept. 16, in A.D. 1895.

"The Sunny Side Company, organized ___ 18, 1897, under the
laws of the State of Connecticut, United States of America, and
legally represented by its president, Hon. Austin Corbin, of New
York, on the one side, and Sebastini Eugenio, agriculturist, and
his family, as undersigned, of Ancona, Italy. Europe on the other
side, stipulate as follows:

"Article I. The Sunny Side Company sells and transfers with
this contract to the above named purchaser, his executor and
legal curators, a lot of ground bearing No. ___ [sic] in the map
hereunto affixed (there was no map affixed) which lot is
described in the map, above referred to, and consists of twelve
and a half acres of ground, already under cultivation, (equal to
5.06 ettari Italian measurements) which makes a part of the Sunny
Side plantation, pertaining as per perfect title of proprietor,
to the selling company, situated in the Chicot county, state of
Arkansas, United States of North America, between the Mississippi
and Lake Chicot, and is one of 250 lots sold contemporaneously to
Italian families with the object of agricultural colonization of
Sunny Side Company's plantation.

---

*Note: this date was printedly incorrectly in the newspaper.

"Art. II. There is included in the sale, the colonial house which is situated on the lot, sufficient for the Sabin_ [sic] family.

"Art. III. The price of the ground and of houses is stipulated at $2,000 in American money, which sum the purchaser obligates himself to pay in American money in the course of twenty-one years to begin from the date of the present contract and payable at dates hereinafter stipulated.

"Art. IV. The purchaser furthermore obligates himself to pay to the Sunny Side Company annual interest at 5 per cent on the balance due until the last payment shall mature as hereinafter stipulated, cancelling his indebtedness.

"Art. V. The first payment will be due in 1896, after the first harvest is made by purchaser upon the ground, and the balance annually in the following manner:

"First year, 1896. For interest on $2000 at 5 per cent, total $100.

"Second year, 1897. Interest on $200 at 5 per cent, plus $50 for reduction of capital, total $150.

"Third year, 1898. For interest on $1950, plus $75 on account of principal; total $172.50.

"Fourth year, 1899. Interest on $1875 at 5 per cent, plus $75 on account of principal; total $178.75.

"Fifth year, 1900. Interest at 5 per cent on $1800, plus $100 on account of principal; total $190.

"Sixth year, 1901. Interest on $1700, plus $100 on account of principal; total $185.

"Seventh year, 1902. Interest at 5 per cent on $1600, plus $100 on account of principal; total $180.

"And for the remaining years is eleven rental payments of $100 each at 5 per cent, as above stated, payable annually at dates of harvest.

"Article VI. For the purpose of furthering the interest of the purchaser, the Sunny Side Company recognizes the ability to accept the payment in part or in full, the said company obligating itself to allow all the time necessary to execute said payments, when on account of some unforeseen cause, the result of the harvest fails to be sufficient to pay the annual rental, it being understood that the payments overdue from one year to another should be paid by purchaser at the first good harvest.

"Art. VII. The Sunny Side Company obligates itself by the present contract to buy, if asked, the cotton that purchaser raises on the property at the current price, quotations as per exchange of the city of New Orleans, state of Louisiana, less the

freight and expenses on said cotton from Sunny Side, said
expenses not to exceed $1 per bale of 500 American pounds, which
will be at the expense of the selling colonist.

"Art. VIII. For the solution of any and all differences that
may arise after the signing of this contract between the purchaser and the Sunny Side Company, the contracting parties agree
to submit to the principle of arbitration, it being stipulated
that the arbitrating commission shall consist of three arbitrators, of which one will be chosen by the selling company, one by
the selling colonist, and the third by the two arbitrators so
named.

"The president of the Sunny Side Company, seller.

2. Cover letter of Mary Grace Quackenbos Report on Sunnyside, September 28, 1907, showing her bold autograph.

Greenville, Miss., September 28, 1907.

The Attorney General,

Washington, D. C.

Sir:

I have the honor to submit to you a report on the Sunny Side plantation, Arkansas. I have discussed at some length the importation of Italian labor and its sequence the "rent contract system"—more rightly called "debt system"—in order to give you a detailed statement of the operations concerning Italian immigrants on cotton plantations in Mississippi and Arkansas. In submitting a full report upon this, the largest plantation, I ask you to regard it as typical of conditions prevailing on the smaller plantations through out the Delta regions. My subsequent reports will be shorter and will supplement matters mentioned in this.

I take this opportunity to mention that in the development of my investigation and work accomplished, I have been materially aided by the excellent work of my three assistants, each of whom has exercised untiring effort to assist me in arriving at the truth. My secretary Hannah Frank undertook to personally interview certain residents in the small towns and has copied from the courts records of cases which have led me to the correct view of the situation. Special Employees, Michele Berardinelli and Charles Pettek have braved a great many dangers; they have been subjected to insults, threatened with arrest, and have been actually driven from certain plantations. You know of the arrest of Employee Pettek at Sunny Side and his fine of $100. in lieu of a three months' chain gang sentence; I have written you from time to time of my own difficulties in entering Sunny Side which eventually was accomplished only through a letter from the Governor of the State. My subsequent visits were made after many and serious controversies and then were merely tolerated but I hope to make other trips to Sunny Side believing that the door to such places should be permanently opened.

With all earnestness I beg to assure you that I am ready for discussion or questions upon each statement I have made.

Respectfully,

Mary Grace Quackenbos.

Special Assistant U. S. Attorney,
Southern District of New York.

3. First page of Mary Grace Quackenbos Report on Sunnyside, September 28, 1907, with a map of Sunnyside by Mary Grace Quackenbos.

SUNNY SIDE COLONY, ARK.
O. B. CRITTENDEN & CO., Lessees,
COTTON FACTORS.

In the southeastern part of the State of Arkansas at a bend of the Mississippi River, a long narrow stretch of land is cut off from the outside world by a circular lake twenty-two miles in length which shapes itself into a horseshoe.

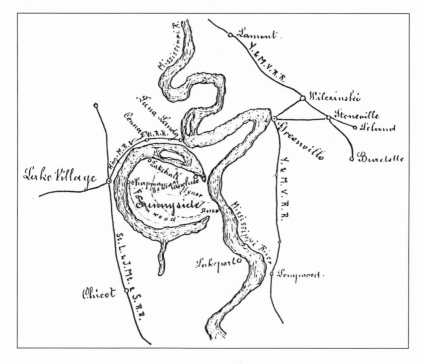

In the bed of the lake there are 16,393 acres of rich soil, — and of the many alluvial lands of the Mississippi delta regions, these are perhaps most famous for their fertility as well as for their poisonous fevers.

Sunny Side is in the eastern portion of this enclosure and is
recognized as the largest Italian "Colony" of the South.
Sometimes it is spoken of as a model of foreign settlements; but
whatever has been said of the prosperity of Sunny Side, it is a
complete failure as an Italian colony. It is simply a huge cotton
plantation divided into four parts — Sunny Side, Hyner, Hebron,
Fawnwood — in which Italians work for American bosses. "The rent
system" is in operation at Sunny Side and "labor"

4. Tenant Contract/1907. English/Italian. Two important differences between the two documents: (1) English version requires an Italian translator; (2) English version requires permission for free sales for debt-free tenants.

Rental Contract.

This contract made and entered into this day, by and between O. B. Crittenden and Company, Landlord, and _____

Tenant, Witnesseth:—

That the said Landlord has this day, and does hereby lease and to farm let, for the year 19____, to said tenant, that certain tract of land comprising_____acres, on the_____ Plantation, in Chicot County, Arkansas, which said tract of land has been pointed out to the said tenant by said landlord, and accepted by said tenant, as pointed out and defined by said landlord, for which said tenant agrees to pay to said landlord for the year 19____, a rental of $_____per acre, aggregating $_____, and the said tenant is to pay all of the indebtedness for monies, supplies, provisions, mules and implements which said landlord may furnish during the year, at prices to be agreed upon, but at current prices if no prices are agreed upon; and to secure the payment of the said rent and supply bill, in addition to the statutory liens provided by law, said landlord is to have a lien, with power of sale, upon the live stock and farming implements of said tenant, described as follows:

_____

_____

_____

and also upon any other live stock and implements which the said tenant may place upon said property during the said year; and in default of the payment of said indebtedness, when due; and in the event of the breach of the covenants and conditions of this contract, the landlord may take immediate possession of all of said property, and may sell it, at public outcry, to the highest bidder, for cash, at the front door of the Court House of Chicot County, or at some point on the leased property, after having first given ten days notice of the time, place and terms of the said sale, by notice posted on said leased

Contratto di Affitto.

Questo contratto redatto e registrato in data di oggi fra O. B. Crittenden and Company, proprietario e_____.

_____

_____fittaiuolo stabilisce quanto segue:—

Che il proprietario ha oggi affittato, per tutto l'anno 190____, a detto fittaiuolo un certo tratto di terra, abbracciante _____ acri, sulla piantagione di cotone, detta Sunny Side, nella Contea di Chicot, stato di Arkansas. Detto tratto di terra é stato mostrato da detto proprietario al detto fittaiuolo e da esso fittaiuolo accettato nello stato descritto e mostrato, da detto proprietario, per il quale detto fittaiuolo accetta di pagare al proprietario per l'anno 19 un affitto di $_____per acre, ammontante a $_____; il detto fittaiuolo dovrá pagare tutti i debiti di denaro, provviste, viveri, muli e strumenti agricoli che il detto proprietario potrebbe fornire durante l'anno, a prezzi da stipulare o a prezzi correnti del mercato, se non stipulati; e per garentirsi del pagamento della pigione e delle provviste il detto proprietario in addizione all'ipoteca stabile accordata dalla legge prenderá ipoteca con potere di vendita, sul bestiame e sugli utensili agricoli del detto fittaiuolo descritti come segue;—

_____

_____

_____

ed anche su tutto il bestiame e strumenti che detto fittaiuolo potrebbe mettere nella proprietá durante il detto anno, e in mancanza di pagamento di detti debiti, quando scaduti; e nel caso di trasgredimento alle clausole e condizioni di questo contratto, il proprietario avrá il diritto di entrare.

property; and, at said sale, the landlord may become the purchaser. Out of the proceeds of the sale, shall be paid the indebtedness due to the landlord, and the balance shall be paid over to the tenant.

The tenant agrees to work the land leased, as directed by the landlord, and under his supervision, and to keep the ditch banks and turn rows weeded, and to gather all of the crops raised on said leased premises and to prepare the same for market. The cotton raised on said leased premises is to be delivered at the landlords gin at Sunny Side, Arkansas.

If the tenant fails or refuses to cultivate the land leased, as directed by the landlord, the landlord shall have the privilege of hiring, at the tenants expense, labor to cultivate said crop, or he shall have the privilege of terminating the lease, and taking possession of the property and crops, the tenant to be responsible for the rent of the land and for the supplies advanced up to the time, for the payment of which the security hereby given can be enforced.

It is agreed that all of the cotton and cotton seed grown by the tenant shall be handled by the landlord, the same to be paid for at current prices, the cotton to be bought by the landlord at prices agreed upon between him and the tenant, but, if the tenant so desires, the cotton is to be sold by O. B. Crittenden and Company, as Cotton Factors, for the account of the tenant, accounts of sale to be rendered to the tenant, and only the ordinary commissions charged other customers are to be charged said tenants. However, when the tenant has paid his entire indebtedness to the landlord, including his rent and supply bill, he shall have the privilege of handling said cotton, after it has been ginned, himself, and dispose of it as he may see fit.

The following rate of charges is agreed upon: The landlord is to gin the cotton, charging therefor 50 cents per hundred pounds of lint cotton, and.................................... for the bagging and ties necessary to wrap each bale; and he is, if desired by the tenant, to haul all cotton to the gin. Where such cotton is hauled by wagon, a charge of..........cents per hundred pounds of lint cotton will be made; and where hauled by railroad, ...cents per

nell' immediato possesso di detta proprietá e potrá venderla all'asta pubblica, al maggiore offerente, per contanti, sulla porta del palazzo di giustizia della Contea di Chicot, o in qualche punto della propietá affittata, dopo di aver prima dati dieci giorni di avviso dell'ora, del luogo e delle condizioni di detta vendita per mezzo di avviso affisso in detta proprietá, il proprietario potrá divenire l'acquirente. Dal ricavo della vendita, sará pagato il debito dovuto al proprietario ed il rimanente sará pagato al fittaiuolo.

Il fittaiuolo si obbliga a coltivare detta terra affittata, come stabilito dal proprietario e sotto la sua sorveglianza di mantenere i solchi, puliti da ogni erbaccia, e di raccogliere la raccolta prodotta sulla terra affittata, e prepararla per il mercato. Il cotone cresciuto su detta terra affittata, dovrá essere consegnato nell'opificio a mulino del proprietario, situato a Sunny Side, Arkansas, per la pulitura del cotone

Se l'affitaiuolo mancasse o si rifiutasse di coltivare la terra affittata come voluto dal proprietario, il proprietario avrá il privilegio d'ingaggiare, a spesa del fittaiuolo stesso, lavoranti per coltivare detta raccolta, oppure esso avrá il privilegio di annullare il contratto e di prendere possesso della proprietá e raccolta, il fittaiuolo rimanendo responsabile per la pigione della terra e per le provviste fornite fino a quella epoca, pel pagamento della quali le garanzie date nel presente atto, entreranno in vigore

É convenuto che tutto il cotone e semi di cotone raccolti dal fittaiuolo saranno maneggiati dal proprietario a prezzi correnti, il cotone sará comprato dal proprietario a prezzi stabiliti fra lui e il fittaiuolo; ma, se il fittaiuolo lo desiderasse, il cotone sará venduto da O. B. Crittenden & Co. quali sensali per conto del fittaiuolo che riceverá i rispettivi conti di vendita e solamente la commissione abituale imposta ad altri clienti sará addebitata al detto fittaiuolo. Peró appena il fittaiuolo avrá pagato tutto il suo debito al proprietario, incluso la pigione e la fattura pelle provviste avrá il privilegio di disporre come crede di detto cotone, dopo esser stato mondato. La rata delle spese convenuta ó la seguente:—

hundred pounds of lint cotton. The charge for hauling the cotton from the gin to the landing is to be _____ cents per bale, and the landing charge on cotton _____ cents per bale.

If the crops on said land have not been gathered at the end of the year, the tenant is to be allowed to gather them, and, for that purpose, is to have ingress to, and egress from, the leased premises, until he shall have the opportunity of gathering them, without paying additional rent therefor, but this right is not to interfere with the rights of the landlord to take possession of the house and premises.

This contract is written in duplicate, and in Italian and English.

Witness the signatures of the landlord and tenant hereto affixed, this _____ day of _____ 190 ___ .

_____
Landlord.

Witness. _____

_____
Tenant.

Witness. _____

STATE OF ARKANSAS.
COUNTY OF CHICOT.

Be it remembered that on this day personally appeared before me, _____ a Justice of the Peace in and for the County aforesaid,

_____

to me well known as the tenant in the foregoing lease, and stated that he had executed the same for the consideration and purposes therein mentioned and set forth.

Witness my hand and official seal as such Justice of the Peace on this the _____ day of _____ 190 ___ .

_____
Justice of the Peace of Chicot, Arkansas.

Il proprietario dovrá pulire il cotone a ragione di 50 cents per ogni cento libbre di cotone greggio e _____ cents per l'imballaggio di ciascuna balla e se desiderato dal fittaiuolo, il proprietario dovrá portare il cotone allo stabilimento. Pel cotone trasportato con carri _____ cents per ogni cento libbre di cotone greggio saranno pagati del fittaiuolo, ese trasportato per ferrovia _____ cents per ogni cento libbre di cotone greggio. Il prezzo pel trasporto del cotone dall'opificio allo scalo é di _____ cents per balla, e le spese di sbarco sono di _____ cents per balla.

Nel caso che la raccolta non fosse stata fatta per la fine dell'anno, una dilazione sará accordata al fittaiuolo per raccoglierla, e a tale uopo gli sará concesso il transito nella terra affittata finché avrá avuto l'opportunitá di completare la raccolta, senza pagare altra pigione, ma questo privilegio non deve interporsi nei diritti del proprietario di prender possesso della casa e terra.

Questo contratto é scritto in duplicato, in Italiano e Inglese.

Testimoni alle firme del proprietario e fittaiuolo qui sottoscritti oggi _____ di _____ 190 ___ .

_____
Proprietario.

Testimoni. _____

_____
Fittaiuolo.

Testimoni. _____

STATO DI ARKANSAS.
CONTEA DI CHICOT.

Si fa noto che oggi personalmente si presentó davanti di me _____ giudice di pace nella Contea _____ a me noto quale fittaiuolo nel precedente contratto, e dichiaró di aver eseguito il medesimo per i termini e le condizioni in esso menzionati ed esposti.

A testimoni di che appongo qui la mia firma e il mio bollo ufficiale quale giudice di pace oggi li _____ di _____ 190 ___ .

_____
Giudice di pace della Contea di Chicot, Arkansas.

For Amendments (over)

5. Amendments to 1907 Contract. Mary Grace Quackenbos negotiated with Percy and Crittenden.

AMENDMENTS.

1.   O. B. Crittenden and Company hereby agree to give the option to the tenant of disposing of his cotton in any one of the three following ways:

I.   The Company will buy the cotton in the seed at current prices. This must be bought at the gin. The tenant will pay the cost of transportation to the gin, or haul it to the gin, as he sees fit.

II.   The Company will buy the cotton after it is ginned and will pay the current price thereof to the tenant.

III.   The Company will sell the cotton for the tenant at the Greenville market realizing the best price for the tenant and charging him only the ordinary commission of two and one half percent (2 1/2%).

If the Company buys the cotton after it has been ginned (II) or sells it at Greenville market (III) the tenant agrees to pay for the ginning, bagging and ties.

2.   The Company agrees that the tenant or the Italian Government shall have a representative at Sunny Side at the gin, or at the store where all of the cotton sales are made, who can keep the tenants posted in regard to the current prices on list and seed cotton and who can advise with the tenants as to the method they shall pursue in disposing of their cotton. The Company agrees that this representative, or another, shall be present in Greenville where the cotton of the tenant is to be sold by the Company, said representative to discuss whether the price at which the cotton is sold is a fair price.

3.   O. B. Crittenden and Company agree that they will properly ceil the tenants' houses and make them weather-proof doing such repairs as will place them in this condition.

4. The Company agrees that an Italian doctor selected by the Italian Government shall practice medicine at Sunny Side, he to charge One Dollar ($1.00) per visit to the tenants. Where the tenants are unable to pay the doctor's fee in cash, the Company agrees to advance the money, charging the tenant 6% interest on such advances. The Company agrees to accept no rebate from the doctor and to pay to the doctor a monthly salary of Twenty-five Dollars ($25.) for actual subsistence.

5. The Company agrees to distribute free of charge to the tenant any quinine furnished by the Italian Consul.

6. The Company agrees that they will permit in Sunny Side two or more Sisters of Mercy or an Italian school teacher, selected by the Italian Government, to live on the Sunny Side property and have free intercourse with tenants for purposes of education, religion, nursing and advice. The Company agrees to furnish a house for said Sisters of Mercy or school teacher.

7. The Company agrees that hereafter only a yearly interest shall be charged and that interest shall never in any instance be reckoned "flat" (which means interest on the amount and not annual interest).

8. The Company agrees to advance to the tenants One Dollar ($1.) per acre for living expenses and to pay to the tenant from said amount twenty five percent (25%) in cash.

9. The Company agrees to employ the tenants, so far as practicable, as day laborers, to work on the railroads, ditching, clearing land and picking cotton other than their own.

10. The Company agrees to make no speculation on medicine sold by the doctor.

11. The Company agrees to have groceries weighed in the presence of the tenant when requested by him.

12. The Company agrees to have each purchase sold to the tenant marked by the storekeeper in a proper book to be kept by the tenant so that he may control the exactness of the monthly bills sold to his account, if the tenant so requests.

13. The Company guarantees that it will use its best efforts to see that its resident managers and clerks at Sunny Side carry out the terms of this contract in honesty, kindness and consider-

ation toward the tenant, and that all matters of controversy brought to the members of the firm will be heard and passed upon in accordance with what is right and just.

Approved and to be adopted
by
O. B. Crittendon & Co.

by Leroy Percy, Atty.
and member of the firm

The Italian physician, and Italian cotton expert mentioned, have been promised by Sig. Feelie Consul at New Orleans, now in Rome, etc. The Italian Government has promised to send the best quality of quinine free of charge. The question of Italian School I have taken up with Mr. Percy the priest and I hope to see the Bishop while in Little Rock. The Corbin heirs will also be requested to deed plot.

6. First page of Mary Grace Quackenbos Report of August 2, 1907, with map of Delta region in Quackenbos's hand.

Greenville, Miss., August 2, 1907

The Attorney General,

Washington, D. C.

Sir:

    I beg to submit herewith a general outline of the location, condition and transgressions found through my investigation in connection with plantations in Mississippi and Arkansas where Italians are at this time working in the cultivation of cotton.

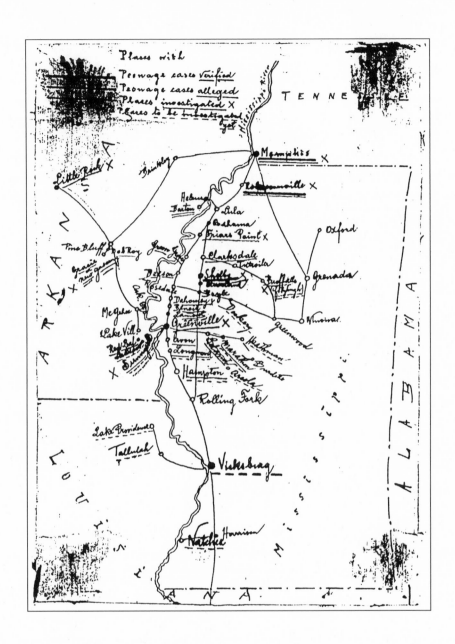

7. Italian tenant receipt from Sunnyside store. Quackenbos has underscored the uniform entry specifying debt owed to the company.

8. Mr. Percy's List. List of tenants who received cash balances in 1906 with Quackenbos's notations (in italic) showing that many had been on Sunnyside upwards of nine to ten years.

MR. PERCY'S LIST.

Amount paid to Italians in cash in crop of 1906 above all supplies and rent charged to them.

Page Ledger	Name	No. Acres.		Amount
1	Susanni Allesandra	36		$198.36
3	Georgini Antonio			344.29
6	Bonchetti Achelli	33 1/8		23.70
8	Alpa Allisandro	20	*9 years*	966.54
9	Catalini Achelli	31	*escaped*	592.77
11	Vignoli Agusto	19		5.66
13	Nandolini Antoni	16		480.68
14	Olivar Alfonso	38 1/4	*12 yrs.*	504.63
15	Rotatori Allesandro	18	*(?)*	428.50
16	Rossini Agusto	25		746.41
17	Silverstreni Allesandro	19 1/2		459.90
18	Pototi Antonio	26 1/2		928.34
8	Luigi Baccusi	25		502.64
11	Cezaro Bonvini	26 1/4		682.44
13	Geovanni Bonvini	18		156.39
5	Serofini Catalini	33 1/4	*10 yrs*	696.81
1	Angelleto Dominico	27		1090.84
3	Antoni Duristanti	19		393.82
7	Vincenzo Duristante	26 1/4		858.74
3	Petro Fratesi	28 1/3	*11 yrs*	555.82
8	Pieroni Fidelli	24 3/4		234.96
11	Serofini Frateni	36	*11 yrs*	277.75
12	Cezari Frateni	31	*11 yrs*	399.69
14	Scucci Francesco	14	*12 yrs*	481.87
15	Belletinni Francesco	11		177.73
16	Greanti Francesco	29		960.42
1	Lamandini Gulelano	26 1/2		262.95
6	Francessini Gulelino	24		545.60
10	Fracossi Guisippi	18		868.61

20	Pusanti Gulelimo	12 1/2		252.86
21	Polidori Giovanni	24 1/2		894.66
6	Picorni Leopoldo	21 1/3		536.29
1	Zampini Mariah	15		12.00
2	Charles Mariangoni	12 1/2		534.22
5	Giovanni Mangorelli	16		107.55
1	Agnoletti Nozarenno	16 3/4		248.14
3	Frateni Nazarenno	28		530.89
9	Cosavocci Nazareno	23 3/4		271.03
12	Coroloni Nazareno	21 1/2		510.92
13	Greganti Nazareno	37 1/4	*9 yrs*	1173.99
14	Siena Nazareno	25	*7 yrs.*	1157.57
15	Gerolomette Nazareno	25		878.67
16	Gulinelli Nazareno	17		810.80
1	Xantini Osualdo	16	*10 yrs.*	176.15
2	Cezarenni Oresto	15 1/4	*7 yrs.*	339.65
4	Manzini Phillipi	15		145.16
5	Giovanni Papelio	15 1/2		313.72
6	Zampi Pietro	23		749.36
7	Gesolini Paolosini	13 1/4	*12 yrs.*	527.04
8	Papelio Satrennia	18 1/2		120.73
10	Alberto Parenti	24 1/2		494.59
11	Eduardo Paolosini	29		565.50
12	Guiseppe Pieroni	19		728.36
2	Reginelli and Co.	30 3/4	*12 yrs.*	201.71
5	Reginelli Rosa	14	*7 yrs.*	344.57
6	Reginelli Antoni	24	*12 yrs.*	692.77
1	Zack Simmonetti	42 1/4		103.20
2	Torigoni Silvini	35		534.79
6	Regnelli Silvini	26 1/2	*8 yrs.*	491.71
7	Maucini Serofini	15		41.46
9	Nicola Stupendo	17		339.60
10	Letrici Silverstreni	23		563.46
13	Eduardo Silverstreni	12 3/4	*(?)*	284.55
14	Giovanni Silverstreni	16 1/2		146.05
15	Fred Sparcorelli	22		740.83
1	Mingorelli Ubaldo	12 1/2		105.01
3	Ludwicio Visconi	23 1/2		413.60
4	Valorio Valbreanni	20		364.20
5	Vifori Andriola	24		693.90
6	Vincenzo Brunetti	26		951.95
	Laniz Zucone		*11 yrs.*	151.00

$34454.02

9. Annual account of Sunnyside tenant Tamboli Guiseppe itemizing expenses, including the "pay roll." From Quackenbos Report of September 28, 1907.

SUNNY SIDE STORE
PLANTING AND GENERAL MERCHANDISE.

Sold to   Sunny Side Arkansas   4/8

M. Tamboli Guiseppe

Feby.	1.	To balance 1905	53.75
	10.	Checks	20.00
Mar.	10.	Pay Roll	1.50
Apr.	7.	Checks (food money)	20.00
May	5.	Checks	20.00
	12.	Mdse.	8.85
June	2.	Pay Roll	12.00
	9.	Checks	20.00
	15.	S. S. Co., 1 mule	150.00
	16.	Pay Roll	16.00
July	3.	Checks	15.00
		Checks	10.00
Aug.	16.	M. M. Norton (doctor)	1.25
Sept.	1.	Checks	15.00
	15.	Priest	.65
	25.	Rent and shopwork	175.35
	26.	M. M. N. (doctor)	3.50
Oct.	6.	Checks	15.00
		Mdse.	2.00
	19.	M. M. N.	7.75
Nov.	2.	Pay Roll	3.50
	9.	Checks	15.00
		Cash	5.00
		M. M. N.	1.25
	27.	Ginnings	7.82
		Hauling	2.91
		Interest	26.70
	28.	Priest	1.00
Dec.	2.	M. M. N.	2.50
Dec.	8.	Checks	2.00
		Gin	7.92

```
                    Hauling                            2.96
                    Interest                           1.95
1907.
Jany.    21.        Pay Roll                          10.25
         26.        Pay Roll                          12.60
         31.        Priest                             1.00
Feby.     2.        Pay Roll                          11.75
         15.        Pay Roll                          14.60
         19.        M. M. Norton                       2.00
                    Hauling Cotton                      .75
Mar.      2.        Pay Roll                            .30
                    M. M. N.                           1.00
                    Ginning                           17.22
                    Hauling                            6.61
                                                     ───────
                                                                   $744.19


Crd.
Nov.     27         Cotton                            93.20
                    Seed                              11.06
Dec.     19.        Cotton                            94.80
                    Seed                              10.07
April     5.        Cotton                           204.00
                    Seed                              22.48          436.60
                                                     ───────
                         Total Debt.   April 1907                   307.59
```

The items added separately show:

Rent $175.00

```
Checks (advances for subsistence)        20.00   Feby. 1906
                                         20.00   April
                                         20.00   May
                                         20.00   June
                                         15.00   July
                                         10.00   July
                                         15.00   Sept.
                 (Food money cut down    15.00   Oct.
                 because mule not working 15.00  Nov.
                 during these months)    15.00   Dec.
                                          2.00   Dec.
                                        ───────
                                        167.00


"Mdse."                                   8.85   May 1906
                                          2.00   Oct.
                                        ───────
                                         10.85   April 1907
```

Interest	26.70	Nov. to Jan./ 07,
	1.95	to April, 1907.
	28.65	
Cash	5.00	Feby./06 to Ap. /07
Ginning Expenses from	7.82	Nov.
Feby.1906 to	7.92	Dec.
April 1907	17.22	Mar.
	26.96	
Hauling Cotton to Gin	2.91	Nov.
	2.96	Dec.
	.75	Feby.
	6.61	Mar.
	13.23	
M. M. Norton (Physician)	1.25	Aug. 1906.
	3.50	Sept.
	7.75	Oct.
	1.25	Nov.
	2.50	Dec.
	2.00	Feby. 1907.
	1.00	Mar.
	19.25	
Priest	.65	Sept. 1906.
	1.00	Nov.
	1.00	Jan. 1907
	2.65	
"Pay Roll"		
(System of putting negroes	1.50	Mar. 1906
to work on the cotton and	12.00	June
making tentant pay their	16.00	June
wages)	3.50	Nov.
	2.00	Dec.
	10.25	Jan. 1907.
	12.60	Jan.
	11.75	Feby.
	14.60	Feby.
	.30	Mar.
	84.50	
Balance debt of year 1905	53.75	

10. Cover letter of Leroy Percy complaint to President
Roosevelt with his autograph. November 14, 1907.

EUROPEAN PLAN   CABLE ADDRESS NEW WILLARD

THE NEW WILLARD,
PENNSYLVANIA AVENUE, FOURTEENTH & F STREETS

WASHINGTON, D.C.   November 14,/07, *190__*

President Theodore Roosevelt,
    White House.

Mr President:
    I enclose the letter about Mrs. Quackenbos which you sug-
gested. I deprecate its length, but could not make the matter
clearer in fewer words, and I trust that you will feel that the
importance of the matter justifies the infliction. As an evidence
of the deep interest felt in this subject by the property holders
in the Delta interested in immigration Mr. J. A. Crawford of
Indianola, one of our largest planters, Mr. Allen Gray of
Evansville, Indiana, operating the largest single plantation in
the South, Mr. Bolton Smith of the firm of Caldwell & Smith of
Memphis controlling a property of 58,000 acres in Louisiana which
they hope to fill with immigrants and Judge Dickinson, general
counsel to the Illinois Central Railroad, all were willing to
come on with me and present this matter to you. I felt satisfied,
however, that upon my bringing the facts to your attention you
would take whatever steps the conditions made advisable.
Commissioner Neill mentioned to me on yesterday that Mrs.
Quackenbos was going South yesterday evening and expected to
remain there several weeks. Of course I said nothing to him in
reference to your remark that you would not go South again.
    If after reading the correspondence you should consider a
further interview with me in any way desirable, I will be glad to

remain here for that purpose, otherwise shall return home
tonight. I mention this in view of the non-committal attitude
taken by General Bonaparte referred to in my letter. Would you
kindly let me hear from you in regard to this at the Willard
today.

Sincerely yours,

11. Last page of Mary Grace Quackenbos's rejoinder to Attorney General Bonaparte concerning Percy's complaint to Roosevelt.

11/18/07 ,No.5

Crittenden and myself entirely out of the question, if immigrants have been writing home to Europe of peonage and the like in the south, and if there are such things there, who are the better friends of the south, this shrewd lawyer who would like to suppress my report and me in a vain endeavor to conceal the facts, or we who are trying to improve the condition and thereby the reputation of the south, to arouse local public opinion there against such ill-treatment of workingmen as exists, and, meantime, to let it be known that the Federal Government has a large force of attorneys and investigators to whom any immigrant molested in such a way can appeal for assistance, with the certainty that his appeal will lead to protection and redress, so far as it is in the power of the Government to afford them?

I heard the President when he made his forceful address at Vicksburg say that one should not be frightened away from his duty and that equal justice must be dealt alike to the rich and to the poor. These words inspire me with the belief that the matter of Mr. Percy's complaints against me, and effort to prejudice my position as your assistant, and avoid the real issues, will be safe in your and the President's hands.

If the President will glance through the annexed list of peonage complaints he will perhaps gain a better idea of the extent and gravity of this peonage work. For reasons which I trust I have made plain I very respectfully ask to be permitted to resume my journey to the south this evening or tomorrow.

Very respectfully,

Mary Grace Quackenbos

Special Assistant.

12. Judge H. C. Niles' instruction to Federal Grand Jury (Crittenden peonage case).

To the Federal Grand Jury now in session at Vicksburg:

I have been requested to give you some directions about the relation to peonage of the State law to pu[r]nish contracting and receiving advances with intent to defraud. In my opinion that law is an attempt to establish what the Constitution and laws of the United States forbid, and is therefore void. What can not be done directly the law says can not be done indirectly; that is, the law refuses to be circumvented, and makes any effort to circumvent unlawful. Freedom from involuntary servitude is every man's right guaranteed by the Constitution and protected by Sections 1990, 5526 and 5525 of the Revised Statutes. A State law calculated to compel a man, by threat of imprisonment, to remain in servitude, and making him prima facie guilty of a crime upon evidence showing only a breach of contract, violates the letter and spirit of the Constitution. In my opinion that is the purpose and effect of this law, or one of its purposes.

Upon suggestion of the District Attorney I also instruct you that the responsibility for law questions is upon him and the Court and that the Jury are to take as the law applicable to the facts before the jury what the District Attorney states as the law, or else submit to me any question of law that the jury may not be satisfied about. Also that no State criminal law, valid or invalid, can be lawfully used to force a debtor by fear of imprisonment to work for another against his will. Also that the kind of force, threats or intimidation unlawful to be used to cause a man to be placed in servitude or to remain there is whatever would naturally prevent a man of the particular kind, under the circumstances of the particular case, from exercising his free will about the matter. An ignorant foreigner in a country whose people and language he does not know, can be intimidated more easily, it may be, than a native well acquainted with the surroundings. The question is whether the particular person is acting upon his free will, or his will has been overcome by fear or force.

Respectfully,

H. C. Niles,

Judge.

Notes

Sunnyside: The Evolution of an Arkansas Plantation, 1840–1945

1. For historical sketches of Chicot County and its cotton plantations see *Biographical and Historical Memoirs of Southern Arkansas* (Chicago: Goodspeed Publishing Co., 1890), 1058–68, J. M. Buffington, "A Brief History of Chicot County in the Vicinity of Lake Village," typescript, 4, Gatewood Branch, Chicot County Library, Lake Village, Arkansas; E. Leona Sumner Brasher, "Chicot County, Arkansas: Pioneer and Present Times," handwritten, Special Collections, Mullins Library, University of Arkansas, Fayetteville; Mrs. W. Garland, "Some Chicot County History," typescript, Arkansas History Commission, Little Rock; Maude Carmichael, "The Plantation System in Arkansas, 1850–1876," unpublished Ph.D. diss., Radcliffe College, 1935, 87–91.

2. Brasher, "Chicot County, Arkansas," 3.

3. Little Rock *Arkansas Gazette*, September 15, 1830, January 25, 1832; Goodspeed's *Southern Arkansas*, 1062; on Abner Johnson's land acquisitions see Chicot County Deed Book A, 247, 248, 412; Deed Book B, 141–43, 394–95; Deed Book C, 133, 146, 433, 434, 469, 471, 498; Deed Book D, 498; Chicot County Deed Records, Books A through F-1 (1823–1886) are on microfilm at the Arkansas History Commission, Little Rock; other deed records referred to herein are located in the Chicot County Courthouse, Lake Village, Arkansas.

4. Chicot County Deed Book C, 556–62.

5. See, for example, the mortgage cited in Chicot County Deed Book D, 450–53.

6. W. H. Perrin, J. H. Battle, and G. C. Kniffer, *Kentucky: A History of the State* (Louisville, Ky.: F.A. Batley Co., 1886), 164; Willard R. Jillison, *Old Kentucky Entries and Deeds* (Louisville, Ky.: Standard Printing Co., 1926), 72; Goodspeed's *Southern Arkansas*, 1088.

7. For the genealogy of the Johnson family and Richard M. Johnson's "mulatto" consort and children, see Leland W. Meyers, *The Life and Times of Colonel Richard M. Johnson of Kentucky* (New York: Columbia University Press, 1932), 13–48, 155, 193, 317–23.

8. *Bowman v. Worthington (1867) Cases Argued in the Supreme Court of the State of Arkansas* (Little Rock: Woodruff and Blocker, 1867), 522–39.

9. Carol S. Jacobs (Oberlin College Archives) to the author, March 20, 1990; Mifflin W. Gibbs, *Shadow and Light: An Autobiography* (New York: Arno Press, 1968), 228–29; Brashner, "Chicot County, Arkansas," 24–25; James W. Mason to General J. W. Sprague, June 23, 1866, Freedmen's Bureau Records, Field Reports, Arkansas (National Archives, Washington, D.C.).

10. Chicot County Deed Book G, 274–75, 300–03, 355–57, 390–91, 490–91, 524–26, 617; Deed Book H, 15–20, 273–75, 283–84, 520–24, 589; Deed Book I, 5–7, 97, 158, 405–06, 416–17, 516–21.

11. Chicot County Deed Book D, 528–34; see also Charles E. Cauthen, ed., *Family Letters of Three Wade Hamptons, 1782–1901* (Columbia, S.C.: University of South Carolina Press, 1953), xv–xvi, 56, 68; *The National Cyclopaedia of American Biography* (New York: James T. White and Co., 1929), VI, 434.

12. Chicot County (Tax Assessment Roll), 1860, 47, Chicot County Court House, Lake Village, Arkansas; Manuscript Census Returns, Eighth Census of the United States, 1860, Chicot County, Arkansas, Agriculture Schedule, 5; Eighth Census of the United States, 1860, Chicot County, Arkansas, Slave Schedule, 194 (Microfilm).

13. Goodpeed's *Southern Arkansas*, 1064–65; see also Kenneth Story, "A Jewel of the Delta: Lakeport Plantation," *Arkansas Preservation*, IX (Winter 1990): 1, 3.

14. Lewis Cecil Gray, *History of Agriculture in the Southern United States to 1860*, 2 vols. (Washington: Carnegie Institution, 1933), II, 1027.

15. O. E. Moore to Joseph Medill, January 29, 1872, in *Arkansas Gazette*, February 4, 1872.

16. Norman E. Clarke, ed., *Warfare Along the Mississippi: The Letters of Lieutenant Colonel George E. Currie* (Mount Pleasant, Mich.: Clarke Historical Collection, 1961), 102.

17. William L. Shea, "Battle of Ditch Bayou," *Arkansas Historical Quarterly*, XXXIX (Autumn 1980): 195–207; see also Clarke, *Warfare Along the Mississippi*, 101–11; *The War of the Rebellion: A Compilation of Official Records of the Union and Confederate Armies* (Washington, D.C.: Government Printing Office, 1880–1901), Series I, XXXIV, Part 1, 971–85.

18. Gavin Wright, *The Political Economy of the Cotton South: Households, Markets and Wealth in the Nineteenth Century* (New York: W. W. Norton, 1978), 89.

19. Application for the Restoration of Property in Chicot County, Arkansas, February 27, 1866, [including a copy of Worthington's presidential pardon], Freedmen's Bureau Records, Field Reports, Arkansas; Special Order No. 27, March 22, 1866, ibid.

20.   Worthington's declining health and financial problems are spelled out in the voluminous unpublished depositions and testimony in the *Worthington v. Mason* case, Circuit Court of the United States, Eastern District of Arkansas, Little Rock, 1875, 1876; according to the census of 1870, Worthington still owned 5,000 acres of land (1,000 acres improved and 4,000 acres unimproved) valued at $60,000; see Manuscript Census Returns, Ninth Census of the United States, 1870, Chicot County, Arkansas, Agriculture Schedule, 2 (Microfilm).

21.   Chicot County Deed Book L, 143–45, 168–69.

22.   Ibid., 170–71, 257–63, 416–17, 421; *Bowman v. Worthington,* 530; *Worthington et al. v. Welch, Administrator, Reports of Cases at Law and in Chancery Argued and Determined in the Supreme Court of the State of Arkansas* (1872) (Fort Smith, Ark.: Calbert McBride Co., 1919), 463–66; Lake Village, (Ark.) *Lake Shore Sentinel* February 11, 1876; *Worthington v. Mason,* 101 U.S., 149-53, 25L., 848; on the movement of slaves to Texas by their owners during the Civil War, see Leon F. Litwack, *Been in the Storm So Long: The Aftermath of Slavery* (New York: Alfred Knopf, 1979), 30–35.

23.   Chicot County Deed Book L, 586.

24.   A. G. Cunningham to William Starling, et al., April 12, 1868, Freedmen's Bureau Records, Field Reports, Arkansas.

25.   On the Chicot "Massacre," see Little Rock *Arkansas Gazette,* December 16, 21, 22, 23, 24, 27, 28, 29, 30, 31, 1871; Chicot County, Circuit Court Records, 1873, 473, Chicot County Court House, Lake Village, Arkansas.

26.   Little Rock *Arkansas Gazette,* October 30, November 2, 1872, February 10, 1874, July 31, 1879.

27.   *The National Cyclopedia of American Biography* (New York: James T. White Co., 1906), XIII, 506–07; ibid., XXXIV, 231; on the movement of blacks into the Delta, see Vernon L. Wharton, *The Negro in Mississippi, 1865–1890* (New York: Harper and Row, 1965), 106–16.

28.   Schedule of Personal Property, Chicot County Deed Book N, 486–88.

29.   Chicot County Deed Book X, 441; Little Rock *Arkansas Gazette,* June 13, 1883; Acts of the General Assembly in the State of Arkansas, 1883 (Little Rock, 1883), 163.

30.   Chicot County Deed Book A-1, 401–2, 417–18, 420–28, 429–33, 434–35, 437–39, 503; "Agreement Between Julia J. Johnson and Lennie Calhoun," Deed Book F-1, 197.

31.   Chicot County Deed Book A–1, 420–28, 477–80; Deed Book B–1, 322, 325–32; Deed Book C–1, 491–95.

32.   New Orleans *Daily Picayune,* March 21, 23, 1882.

33.   Ibid., March 23, 1882.

34.   *Senate Report of the Committee of the Senate Upon the Relations of Labor and Capital and Testimony Taken by the Committee,* 6 vols.

(Washington, D.C.: Government Printing Office, 1885), II, 157–59; hereinafter cited as *Senate Committee Report and Testimony*.

35. Ibid., 168–69.

36. Ibid., 173, 177–78.

37. Ibid., 170, 173, 176, 187.

38. Ibid., 173–75.

39. Timothy Thomas Fortune, *Black and White: Land, Labor and Politics in the South* (New York, Arno Press, 1968), 178–79, 242–89.

40. *Senate Committee Report and Testimony*, 175–76.

41. Ibid., 159–60.

42. Ibid.; see also Chicot County Deed Book C-1, 233, 399–400, 473, 541.

43. New Orleans *Daily Picayune*, March 26, 1882.

44. Little Rock *Arkansas Gazette*, December 31, 1886.

45. *New York Times*, December 19, 1918; Chicot County Deed Book B-1, 322, 325–32; Chicot County Deed Book C-1, 487–90.

46. J. M. Rose, Commissioner of United States Circuit Court, to Gilmour S. Moulton, October 26, 1885, filed in Chicot County Deed Book F-1, 573–87.

47. Ibid., 563–75.

48. Maury Klein, *The Great Richmond Terminal: A Study in Businessmen and Business Strategy* (Charlottesville, Va.: University Press of Virginia, 1970), see especially 32–34; on the later career of John C. Calhoun, see Marion J. Verdery, "The Southern Society of New York," *National Magazine* XV (1892): 547–50.

49. *The National Cyclopaedia of American Biography*, XXXI, 279; Chicot County Deed Book I-1, 221–37; ibid. M-1, 361–72; on Corbin's anti-Jewish activities, see Pine Bluff (Ark.) *Weekly Press*, July 31, August 7, 1879; *New York Times*, July 23, 1878; *New York Tribune*, August 20, 1879.

50. Ed Trice, "Shadows over Sunnyside," Federal Writers' Project, Works Progress Administration manuscript, 1, Arkansas History Commission, Little Rock.

51. *Biennial Report of the State Penitentiary of Arkansas for the years 1893 and 1894* (Morrilton, Ark: Pilot Printing Co., n.d.), 68; Hiram W. Ford, "A History of the Arkansas Penitentiary to 1900," unpublished M.A. thesis, University of Arkansas, 1936, 137; Osceola (Ark.) *Times*, February 2, 1901.

52. *Ashley County Weekly Eagle*, (Hamburg, Ark.), August 8, 27, December 10, 1895.

53. Trice, "Shadows Over Sunnyside," 1; Lake Village (Ark.) *Chicot Spectator*, June 26, 1936; *New York Times*, August 28, 1886.

54. *Senate Reports*, 61 Cong., 2 Sess., No. 633, *Reports of the Immigration Commission* 2 vols. (Washington, D.C., 1911), I, 319–20, hereinafter cited *Reports of the Immigration Commission*.

55. *New York Times*, January 11, 1895.

56. New Orleans *Daily Picayune*, December 1, 1895; for an English translation of the standard contract between the Sunnyside Company and the Italian immigrants, see ibid., November 30, 1895.

57. Ibid., December 3, 1895.

58. Shannon Craig, "Arkansas and Foreign Immigration, 1890–1915," unpublished M.A. thesis, University of Arkansas, 1979, 15–32; Little Rock *Arkansas Gazette*, December 10, 1895; *Ashley County Weekly Eagle*, June 11, 1896; James W. Leslie, "John Mack Gracie," *Jefferson County Historical Quarterly*, VII (1978): 17–20; Louis R. Harlan, *The Booker T. Washington Papers*, 13 vols. (Urbana, Ill.: University of Illinois Press, 1972–1984), X, 363–64.

59. *Ashley County Weekly Eagle*, February 20, March 19, October 15, 22, 1896.

60. Ibid., November 18, December 10, 1896.

61. Ibid., May 30, 1897, May 8, 1902; Y. W. Etheridge, *History of Ashley County, Arkansas* (Fort Smith, Ark., 1959), 44–46; the Sunnyside Company granted a right-of-way for the Mississippi River, Hamburg and Western Railway, see Chicot County Deed Book S-1, 450.

62. *Reports of the Immigration Commission*, I, 319–20; Craig, "Arkansas and Foreign Immigration," 61–66; see also Thomas Rothrock, "The Story of Tontitown, Arkansas," *Arkansas Historical Quarterly*, XVI (Spring 1957): 84–88; for the detailed leases with Grayden Drew and with Hawkins, Crittenden, and Percy, see Chicot County Deed Book S-1, 130–35, 158–64.

63. Robert L. Brandfon, "The End of Immigration to the Cotton Fields," *Mississippi Valley Historical Review*, L (March 1964): 591–611; Lee J. Langley, "Italians in the Cotton Fields," *Manufacturers' Record*, XLV (April 7, 1904): 250; Alfred Holt Stone, "Italian Cotton Growers in Arkansas," *Review of Reviews*, XXXV (February 1907): 209–13; Lewis Baker, *The Percys of Mississippi: Politics and Literature in the New South* (Baton Rouge: Louisiana State University Press, 1983), 27–31; for the Sunnyside Company's bond issue of $400,000 in 1910, see Chicot County Deed Records, M-2, 246–51.

64. Chicot County Deed Book E-3, 610; Chicot County Deed Book V-3, 377; Chicot County Deed Book U-3, 278; Chicot County Deed Records, Miscellaneous Book #2, 97; the court order directing the sale of Sunnyside Company land as well as the consummation of the sale is found in Chicot County Deed Book K-9, 17–27.

65. Trice, "Shadows Over Sunnyside," 3.

66. Ibid., 1–3.

67. Originally funded by the Federal Emergency Relief Administration, the Arkansas Rural Rehabilitation Corporation ultimately became a part of the Resettlement Administration; see Michael H. Mehlman, "The Resettlement Administration and the Problems of Tenant Farmers in Arkansas, 1935-1936," unpublished Ph.D. diss., New York University, 1970;

on Chicot Farms see *Report of the Administrator of the Farm Security Administration, 1937* (Washington, D.C.: Government Printing Office, 1939), 33; Donald Holley, *Uncle Sam's Farmers: The New Deal Communities in the Lower Mississippi Valley* (Urbana, Ill.: University of Illinois Press, 1975), 114, 284; Chicot County Deed Book B-4, 32–33; ibid. A-5, 71, 261, 331, 416; ibid. D-5, 532; ibid. G-5, 37, 90; ibid. I-5, 206, 209.

68. Trice, "Shadows Over Sunnyside," 1.

Labor Relations and the Evolving Plantation: The Case of Sunnyside

1. The Arkansas legislature, like legislatures in other southern states, defined the lien laws in a manner that worked to the advantage of planters. Merchants, in fact, found their liens subordinate to those of planters. But sharecroppers, and to a lesser extent, the tenant farmers, were especially disadvantaged by certain acts of the Arkansas legislature. *Acts of Arkansas*, 1868, 245; Ibid., 1875, 85. Ibid., 1885, 225–26. Ibid., March 7, 1893, 75–76; Ibid., March 27, 1893; Ibid., April 7, 1893. See also Jeannie Whayne, "Creation of a Plantation System in the Arkansas Delta in the Twentieth Century," *Agricultural History* 66 (Winter 1992): 73–74.

2. For an overview of the few historians who departed from the preoccupation with political rather an economic questions, see Harold D. Woodman, "Sequel to Slavery: The New History Views the Postbellum South," *Journal of Southern History* 43 (November 1977), n. 1.

3. William A. Dunning, *Reconstruction, Political and Economic, 1865–1877* (New York and London: Harper, 1907); Ulrich B. Phillips, "The Central Theme of Southern History," *American Historical Review* 24 (October, 1928): 30–43; Walter Lynwood Fleming, *Civil War and Reconstruction in Alabama* (New York: Columbia University Press, 1905); Charles Ramsdell, *Reconstruction in Texas* (New York: Columbia University Press, 1910); Thomas S. Staples, *Reconstruction in Arkansas* (New York: Columbia University Press, 1923); David Yancy Thomas, *Arkansas in War and Reconstruction, 1861–1974* (Little Rock: Ark. Division, United Daughters of the Confederacy, 1926).

4. Charles Beard, *The Rise of American Civilization* (New York: The Macmillan Company, 1927), 52–55. See also Eugene D. Genovese, "Charles Beard and the Economic Interpretation of History," in *Charles Beard: An Observance of the Centennial of His Birth*, M. Swanson, ed. (Greencastle, Ind.: Dupauw University Press, 1976), 25–44. See also Alrutheus Ambush Taylor, *The Negro in the Reconstruction of Virginia* (Washington, D.C.: The Association for the Study of Negro Life and History, 1926); Carter Woodson,

The Rural Negro (New York, The Association for the Study of Negro Life and History, 1930); W. E. B. DuBois, *Black Reconstruction in America* (New York: Russell and Russell, 1935).

5. C. Vann Woodward, *Reunion and Reaction: The Compromise of 1877 and the End of Reconstruction* (New York: Little, Brown & Company, 1951). C. Vann Woodward, *Origins of the New South: 1877–1913* (Baton Rouge: Louisiana State University Press, 1951).

6. Woodward, *Reunion and Reaction*, 2.

7. For an excellent study of the transformation of the Republican party, see David Montgomery, *Beyond Equality: Labor and the Radical Republicans, 1866–1872* (New York: Alfred A. Knopf, 1967).

8. A number of historians have followed Woodward with excellent studies that examined the Republicans and their goals: Kenneth Stampp, *The Era of Reconstruction, 1865–1877* (New York: Knopf, 1965); W. McKee Evans, *Ballots and Fence Rails: Reconstruction on the Lower Cape Fear* (New York: W. W. Norton & Company, 1966); Michael Perman, *Reunion Without Compromise: The South and Reconstruction: 1865–1868* (New York: Cambridge University Press, 1973); Otto H. Olsen, ed., *Reconstruction and Redemption in the South* (Baton Rouge: Louisiana State University Press, 1980). Others have concentrated on why Reconstruction failed: Thomas Holt, *Black Over White: Negro Political Leadership in South Carolina during Reconstruction* (Urbana: University of Illinois Press, 1977); William Gillette, *Retreat from Reconstruction, 1869–1879* (Baton Rouge: Louisiana State University Press, 1979). For an excellent look at the Woodward thesis in Arkansas, see John William Graves, *Town and Country: Race Relations in an Urban-Rural Context* (Fayetteville: University of Arkansas Press, 1990).

9. Woodward, *Origins of the New South*, 20.

10. James L. Roark, *Master's Without Slaves: Southern Planters in the Civil War and Reconstruction* (New York: W. W. Norton & Company, 1977).

11. Jonathan M. Wiener, *Social Origins of the New South: Alabama, 1860–1885* (Baton Rouge: Louisiana State University Press, 1978). See also Roger L. Ransom and Richard Sutch, *One Kind of Freedom: The Economic Consequences of Emancipation* (Cambridge: Cambridge University Press, 1977).

12. Information conveyed to the author by Willard Gatewood and gleaned from documents in his possession.

13. Willie Lee Rose, *Rehearsal for Reconstruction: The Port Royal Experiment* (New York: Vintage Books, 1964).

14. Louis S. Gerteis, *From Contraband to Freedman: Federal Policy Toward Southern Blacks 1861–1865* (Westport, Conn.: Greenwood Press, 1973).

15. Eric Foner's work demonstrates that planters reasserted their political power in an effort to redefine class relations between the old masters and the

freed slaves. Their continued ownership of the means of production (the plantations) insured their political dominance, and the two factors worked together to overcome the freedmen. Eric Foner, *Nothing But Freedom: Emancipation and Its Legacy* (Baton Rouge: Louisiana State University Press, 1983). Foner pointed out, however, that the South's experience with emancipation was distinguished from that of other slave societies undergoing the same transformation by the fact that freedmen were given political rights. This occurred nowhere else.

16.   Joel Williamson, *The Crucible of Race: Black-White Relations in the American South since Emancipation* (New York and Oxford: Oxford University Press, 1984).

17.   Gatewood, this volume, 12.

18.   Jay R. Mandle, *The Roots of Black Poverty: The Southern Plantation Economy after the Civil War* (Durham: Duke University Press, 1978).

19.   Harold Woodman, "Post–Civil War Agriculture and the Law," *Agricultural History* 53 (January 1979): 319–37.

20.   Pete Daniel, *The Shadow of Slavery: Peonage in the South, 1901–1969* (Urbana and Chicago: University of Illinois Press, 1972).

21.   Jonathan M. Wiener, "Class Structure and Economic Development in the American South, 1865–1955," *American Historical Review* 84 (October 1979): 973.

22.   Stephen J. DeCanio, *Agriculture in the Postbellum South: The Economics of Production and Supply* (Cambridge and London: MIT Press, 1974).

23.   For a stunning refutation of both Higgs and DeCanio, see Harold D. Woodman, "Sequel to Slavery: The New History Views the Postbellum South," *The Journal of Southern History* 43 (November 1977): 523–54.

24.   Gavin Wright, *Old South, New South: Revolutions in the Southern Economy since the Civil War,* (New York: Basic Books, Inc., 1986).

25.   Dwight B. Billings, Jr., *Planters and the Making of a 'New South': Class, Politics, and Development in North Carolina, 1865–1900* (Chapel Hill: University of North Carolina Press, 1979).

26.   Elizabeth Fox-Genovese and Eugene Genovese, *Fruits of Merchant Capital: Slavery and Bourgeois Property in the Rise and Expansion of Capitalism* (New York and Oxford: Oxford University Press, 1983), 398.

27.   Eugene D. Genovese, *Roll, Jordan, Roll: The World the Slaves Made* (New York: Vintage Books, 1972).

28.   Fox-Genovese and Genovese, *Fruits of Merchant Capital,* 389–99.

29.   Mandle, *Roots of Black Poverty,* 30, 32.

30.   Williamson, *The Crucible of Race,* 82, 91, 435–37. While Williamson found paternalism emerging in mill villages, Lawrence Powell found Yankee planters (described as men who came South during and after the Civil War to

prove that blacks could work as free laborers) adopting "yankee paternalism." Powell did not define what he meant by yankee paternalism, but he argued that "once the freedmen showed they were able to use the market to further their own interests . . . yankee paternalism would degenerate into a racism every bit as spiteful and vindictive as that of the old master and maybe more so." Lawrence N. Powell, *New Masters: Northern Planters during the Civil War and Reconstruction* (New Haven: Yale University Press, 1980).

31. Genovese and Fox-Genovese, 398, 389–99.

32. Gatewood, this volume, 16.

33. C. Vann Woodward, *The Strange Career of Jim Crow* (New York, Oxford University Press, 1955).

34. Williamson, *The Crucible of Race.*

35. Loren Schweninger, *Black Property Owners in the South, 1790–1915* (Champaign: University of Illinois Press, 1990).

36. Steven Hahn, "African-American Life in the Nineteenth-Century South: A Review Essay." *Arkansas Historical Quarterly* 50 (Winter 1991): 361.

37. Hahn, "African-American Life in the Nineteenth-Century South," 359.

38. J. Morgan Kouser, *The Shaping of Southern Politics, Suffrage Restriction and Establishing of the One-Party South, 1889–1910* (New Haven: Yale University Press, 1974).

39. Barbara J. Fields, "Ideology and Race in American History," in J. Morgan Kousser and James M. McPherson, eds., *Region, Race and Reconstruction: Essays in Honor of C. Vann Woodward* (New York and Oxford: Oxford University Press, 1982).

40. Cohen, *At Freedom's Edge,* xiv.

41. Steven Hahn challenges Cohen on this very point and suggests that one look to three books in particular for studies that focus on class conflict as important in the formation of the Jim Crow South: Woodward, *Origins of the New South,* 321–95; Kousser, *The Shaping of Southern Politics*; and Steven Hahn, *The Roots of Southern Populism: Yeoman Farmers and the Transformation of the Georgia Upcountry, 1850–1890* (New York: Oxford University Press, 1983).

Peonage at Sunnyside and the Reaction of the Italian Government

1. Ernesto R. Milani, "Marchigiani and Veneti on Sunnyside Plantation," in Rudolph J. Vecoli, ed., *Italian Immigrants in Rural and Small Town America* (New York: American Italian Historical Association, 1987), 18–30; Alessandro Oldrini to Saverio Fava, September 7 and September 12,

1894, in "I Fondi Archivistici della Legazione Sarda e delle Rappresentanze Diplomatiche negli U.S.A., 1848–1901," Archivio Storico del Ministero degli Affari Esteri, I, busta 110, fasc. 2193 (Roma, Istituto Poligrafico e Zecca dello Stato, 1988), hereinafter cited as A.S.M.A.E.

2. Saverio Fava to Regio Ministero degli Affair Esteri, October 10, 1894, A.S.M.A.E.

3. Alessandro Oldrini to Saverio Fava, December 7, 1895, A.S.M.A.E.

4. Alessandro Oldrini to Saverio Fava, November 29, 1895, December 9 and 12, 1895, and Pietro Bandini to Saverio Fava, January 20, 1896, and April 13, 1896, ibid.

5. Regio Ministero degli Affari Esteri to Saverio Fava, July 23, 1896, October 31, 1896; and Saverio Fava to Regio Ministero degli Affari Esteri, January 8, 1897, ibid.; Risultato Conto di Ogni Colono per l'Annata 1896–1897 a Sunnyside, Arkansas, ibid.

6. Giovanni D'Elpidio to Saverio Fava, May 25, 1897, ibid.

7. Giulio Cesare Vinci to Pietro Bandini, August 8, 1897, and Regio Ministero degli Affari Esteri to Saverio Fava, July 12, 1897, A.S.M.A.E., busta 110, fasc. 2194.

8. Pietro Bandini to G. C. Vinci, December 1, 1897, ibid.

9. Egisto Polmonari to Italian Consul at New Orleans, December 4, 1897, and Camillo Romano to G. C. Vinci, December 30, 31, 1897, ibid.

10. See, for example, the New York Sun, January 21, 1898, clipping, ibid.

11. Regio Ministero degli Affari Esteri to G. C. Vinci, January 25, 1898; Guido Rossati to G. C. Vinci, March 8, 1898; "Ai Coloni Italiani di Sunnyside," February 1, 1898, leaflet, ibid.

12. Pietro Bandini to Guido Rossati, March 8, 1898; Austin Corbin, Jr., to G. C. Vinci, March 8, 1898; G. C. Vinci to Regio Ministero degli Affari Esteri, March 10, 1898; Pietro Bandini to G. C. Vinci, March 14, 1898; Alessandro Oldrini to G. C. Vinci, March 30, 1898; and George Edgell to G. C. Vinci, April 22, 1898, ibid.

13. Regio Ministero degli Affari Esteri to Regia Ambasciata a Washington, February 25, 1899, ibid.; G. Saint Martin, "Gli Italiani nel Distretto Consolare di Nuova Orleans," Bollettino dell'Emigrazione, 1903, No. 1, 3–23.

14. Edmondo Mayor Des Planches, Attraverso gli Stati Uniti. Per l'Emigrazione Italiana (Torino, Italy: Unione Tipografica-Editrice Torinese, 1913), 137, 321.

15. Lionello Scelsi, "Relazione del Regio Console in Nuova Orleans circa le Condizioni degli Emigrati Italiani in alcune Localita' di quel Distretto Consolare (Allegato al Rendiconto Sommario dell'Adunanza del 13 Dicembre 1907)," Bollettino dell'Emigrazione, 1908, No. 8, 38–44.

16. Luigi Villari, "Gli Italiani nel Sud degli Stati Uniti," Bollettino dell'Emigrazione, 1907, No. 10, 39–49; Luigi Villari, "Gli Italiani nel Distretto

Consolare di Nuova Orleans" (Stati Uniti), *Bollettino dell'Emigrazione*, 1907, No. 20, 3–46.

17. Scelsi, 38–44; Lionello Scelsi to Augusto Catalani, March 19, 1907, and Leroy Percy to George Edgell, April 2, 1907, Percy Family Papers, Mississippi Department of Archives and History, Jackson, Mississippi.

18. Edmondo Mayor Des Planches to Secretary of State Elihu Root, June 4, 1907, Department of State General Records, Record Group (hereinafter referred to as R.G.) 59, 866.55/6923, National Archives, Washington, D.C.; Mary Grace Quackenbos, "Report on Sunnyside Plantation," September 28, 1907, Department of Justice Records, R.G. 60, 100937, National Archives, Washington, D.C.

19. "Censimento delle Famiglie Italiane nelle Piantagioni di Cotone della Vallata del Fiume Mississippi," *Bollettino dell'Emigrazione*, 1913, No. 5, 195; "Dati circa i Raccolti del 1912 e le Condizioni dei nostri Agricoltori nel Distretto Consolare di Nuova Orleans," *Bollettino dell'Emigrazione*, 1913, No. 7, 97–104; Luigi Villari, "L'opinione Pubblica Americana e i nostri Emigrati," *Nuova Antologia*, CIIL, fasc. 927, August 1, 1910, 497–517; Luigi Villari, "L'Emigrazione Italiana negli Stati Uniti d'America," *Nuova Antologia*, CXXXXIII, fasc. 906, September 16, 1909, 294–311.

Mary Grace Quackenbos and the Federal Campaign against Peonage

1. Edmondo Mayor des Planches to Secretary of State, June 4, 1907, Department of State General Records, Record Group (hereinafter cited as R.G.) 59, 866.55 6923/8–9, National Archives, Washington, D.C.

2. On the demise of the original colony, see Ernesto Milani, "Marchigiani and Veneti on Sunnyside Plantation," in Rudolph J. Vecoli, ed., *Italian Immigrants in Rural and Small Town America* (New York: American Italian Historical Association, 1987), 18–30; on the claim that O. B. Crittenden was one of the most wealthy men of the South, see the *Vicksburg Herald*, October 26, 1907; also, the report of Italian consular agent G. Rosati to Conte G. C. Vinci, March 8, 1898, Italian National Archives, Rome (copy in possession of the author).

3. Lee J. Longley, "Italians in the Cotton Fields," *Manufacturer's Record* 45 (April 7, 1904): 250; Walter Fleming, "Immigration to the Southern States," *Political Science Quarterly* 20 (June 1905): 292–93; Alfred Holt Stone, "The Italian Cotton Grower, The Negro's Problem," *South Atlantic Quarterly* 4 (January 1905): 42–47; Alfred Holt Stone, "Italian Cotton Growers in Arkansas," *Review of Reviews* 35 (February 1907): 209–13; and especially, Alfred Holt Stone, "The Economic Future of the American Negro," a paper

read to the American Economic Association, New York, 1905, and published as chapter 5 in Alfred Holt Stone, *Studies in the American Race Problem* (New York: Doubleday, Page & Co., 1908). For assessments of the efforts to recruit Italians to the Delta, see Robert L. Brandfon, *Cotton Kingdom of the New South: A History of the Yazoo-Mississippi Delta from Reconstruction to the 20th Century* (Cambridge, Mass.: Howard University Press, 1967); and Lewis Baker, *The Percys of Mississippi: Politics and Literature in the New South* (Baton Rouge, 1983). Baker covers the Quackenbos investigation of Sunnyside, but he does not make use of the federal records at the National Archives where Quackenbos's side of the controversy is to be found. For Mary Grace Quackenbos's estimate of the number of Delta plantations employing Italians, see Mary Grace Quackenbos to the attorney general, August 2, 1907, Department of Justice Records, R.G. 60, 100937, National Archives, Washington, D.C.

4. On the heavy rains, see Leroy Percy to W. P. Brown, July 8, 1907, Percy Family Papers, Mississippi Department of Archives and History, Jackson, Mississippi, hereinafter cited as Percy Papers; on letters of complaint, see Des Planches to the secretary of state, June 4, 1907, R.G. 59, 866.55 6923/8–9; on closed immigration to the Delta, see Leroy Percy to Captain J. S. McNeilly, April 3, 1907, Leroy Percy to John M. Parkers, April 5, 1907, and Leroy Percy to W. A. Percy, April 19, 1907, in Percy Papers; on the barn burning, see Leroy Percy to George Edgell, December 19, 1906, Percy Papers; on absconding families, see Leroy Percy to George Edgell, March 9, 1907, Percy Papers.

5. Leroy Percy to George Edgell, February 14, 1907, Percy Papers; Mary Grace Quackenbos, "Report on Sunnyside Plantation," September 28, 1907, Department of Justice Records, R.G. 60, 100937, National Archives, Washington, D.C., hereinafter cited as Quackenbos Report, September 28, 1907.

6. See Stone, "Italian Cotton Growers in Arkansas," 209–13.

7. Mary Grace Quackenbos, "Report on general conditions of Delta cotton plantations," January 10, 1908, R.G. 59, 866.55/8–9, hereinafter cited as Quackenbos Report, January 10, 1908.

8. "Flat" interest charges carried a full year's 10 percent charge without prorating from the time the credit was actually used before annual reckoning. Thomas J. Woofter, who argues that flat interest charges were common on southern plantations, calculates that 10 percent flat interest equates on average to an effective interest of 40 percent per annum; see Thomas J. Woofter, *Landlord and Tenant on the Cotton Plantation* (Washington, D.C.: Works Progress Administration, 1936), 53ff. Of course, the measure of usury worsens the later in the year advances are taken, and this would vary from one tenant to another. For an argument that year-end advances were generally in heavy demand in the cotton economy, see Thomas D. Clark, "The Furnishing and Supply System in Southern Agriculture Since 1865," *Journal of Southern*

History 12 (February 1946): 31–32. Contrary to Woofter and Clark, Harold D. Woodman argues that flat interest charges were *not* the norm on cotton plantations; see Harold D. Woodman, *King Cotton and His Retainers: Financing and Marketing the Cotton Crop of the South, 1800–1925* (Lexington, Ky., 1968; paperback edition, Columbia, S.C., 1990), 56, 36n.

9. Quackenbos Report, September 27, 1907, notes that of the seventy families interviewed during Quackenbos's investigation, fifty-nine carried over debts from the previous year.

10. Quackenbos Report, September 27, 1907, remarks on the deep disillusion among the Italians on Sunnyside.

11. For Quackenbos's work with the Department of Justice in connection with peonage prosecutions, see Pete Daniel, *The Shadow of Slavery: Peonage in the South, 1901–1969* (Urbana and Chicago: University of Illinois Press, 1972), especially chapter 5.

12. Jerrell H. Shofner, "Mary Grace Quackenbos, A Visitor Florida Did Not Want," *Florida Historical Quarterly* 58 (January 1980): 273–90; also, Daniel, *Shadow of Slavery*, chapter 5.

13. *Annual Reports of the Legal Aid Society of New York* from 1899 through 1913, New York Public Library, provide detailed breakdowns of the types of cases handled each year. Cases involving wages and employment disputes far outnumber all other types of cases each year. Furthermore, in her report of the network of New York labor agents involved in filling contracts for southern states, Quackenbos notes of one: "I recognized Schwarts as the most corrupt labor agent on the lower East Side—one whom, in my work among the poor of New York, I had occasion to sue many times." Mary Grace Quackenbos to the attorney general, February 8, 1907, R.G. 60, 100937.

14. Mary Grace Quackenbos to the attorney general, July 20, 1907, Department of Justice Records, R.G. 60, 74682, enclosures, National Archives, Washington, D.C.

15. Ibid.

16. Leroy Percy to Bonaparte, August 19, 1907, R.G. 60, file 100937.

17. Mary Grace Quackenbos, Report to the attorney general of August 2, 1907, R.G. 60, 100937; hereinafter cited as Quackenbos Report, August 2, 1907.

18. Quackenbos Reports of August 2, 1907, and September 28, 1907.

19. Quackenbos Reports of September 28, 1909, and January 10, 1908. Rupert B. Vance argues that the "pay roll" was not an uncommon feature of cotton plantation contracts in the twentieth-century South; see Vance, *Human Factors in Cotton Production* (Chapel Hill: University of North Carolina Press, 1929), 168.

20. Mary Grace Quackenbos to the attorney general, July 20, 1907, R.G. 60, 74683, enclosures, National Archives, Washington, D.C.; Quackenbos Report, September 28, 1907; see also Leroy Percy to Austin Corbin [Jr.],

March 17, 1906, Percy Papers, responding to a rumor Corbin had reported about abusive treatment of Italian tenants by Sunnyside managers.

21. Quackenbos Reports of August 2, 1907, September 28, 1907, and January 10, 1908. In addition to these, Quackenbos wrote a lengthy report condemning labor conditions and noting violations of alien contract labor laws at a cotton mill located near Helena, Arkansas: "The Premier Cotton Mills, Barton, Arkansas," October 10, 1907, R.G. 60, 100937.

22. The negotiations over the company's business practices are detailed in Mary Grace Quackenbos to Leroy Percy, Esq., August 17, 1907, and Leroy Percy to Mrs. Mary Grace Quackenbos, August 17, 1907, R.G. 60, 100937. Whether or not the company implemented these provisions in 1908 is questionable. Correspondence between Leroy Percy and the Sunnyside bookkeeper shows that flat interest was terminated only because the Italians persisted in complaining about it, and the company realized they were violating the law by adhering to it. John Holland to Leroy Percy, October 11, 1907, and Leroy Percy to J. E. Holland, October 15, 1907, Percy Papers. The policy of planters insisting on the right to the tenants' cotton crop instead of allowing free sales seems to have been a standard in tenants' contracts throughout the South; although as Harold D. Woodman points out, such arrangements were not compelled by state lien laws. Planters simply arrogated the right to the finished crop. See Harold D. Woodman, "Postbellum Social Change and Its Effects on Marketing the South's Cotton Crop," *Agricultural History* 56 (January 1982): 215–30.

23. Mary Grace Quackenbos to the attorney general, August 12, 1907, R.G. 60, 100937; Quackenbos Reports of August 2, 1907, and January 10, 1908. Quackenbos to Bonaparte, August 14, 1907, Bonaparte to Quackenbos, August 20, 1907, R.G. 60, File 100937 and Bonaparte to Roosevelt, September 10, 1907, Charles Bonaparte Papers, Library of Congress.

24. Mary Grace Quackenbos to the attorney general, August 2, 1907, R.G. 60, File 100937. On the request of the subpoenas, see, Quackenbos to the attorney general, August 1, 1907, and Henry Louis Stimson to Bonaparte, August 14, 1907. On Quackenbos's earliest clue about peonage on Sunnyside, Charles Wells Russell, *Report on Peonage* (Washington, 1908), 15.

25. On the request of the subpoenas, see Quackenbos to the attorney general, August 1, 1907, and Henry Louis Stimson to Bonaparte, August 14, 1907. Mary Grace Quackenbos to Charles Wells Russell, October 8, 1907, and Mary Grace Quackenbos to the attorney general, October 25, 1907, R.G. 60, 100937; Quackenbos Report, September 28, 1907. Percy's private correspondence reveals two other instances where attempts were made to force absconding Italians back to Sunnyside in the spring of 1907; see Leroy Percy to James B. Yerger, March 11, 1907, Percy to Will Dockery, March 6, 1907, and Percy to Dockery, March 20, 1907, Percy Papers.

26. Leroy Percy to the attorney general, August 19, 1907, and the attorney general to Leroy Percy, August 23, 1907, R.G. 60, 100937.

27. Vicksburg *Herald*, September 8, 1907, and Greenville *Times*, September 29, 1907, both enclosures in Mary Grace Quackenbos to the attorney general, October 25, 1907, R.G. 60, 100937.

28. Mary Grace Quackenbos to the attorney general, October 8, 1907, R.G. 60, 100937; Vicksburg *Herald*, October 26, 1907; Mary Grace Quackenbos to the attorney general, October 25, 1907, R.G. 60, 100937. On the eve of the federal grand jury hearing, J. S. McNeilly's Vicksburg *Herald* rang out with even more strident denunciations, calling Quackenbos a "crack-brained socialist woman," Assistant Attorney General Charles Wells Russell a "Southern scalawag," and Attorney General Bonaparte, "the real villain of the Play"; see Vicksburg *Herald*, January 6, 1908.

29. Mary Grace Quackenbos to the attorney general, October 8, 1907, R.G. 60, 100937; on the labor agent's confession, see Mary Grace Quackenbos to the attorney general, September 21, 1907, ibid.; the agent's deposition to Quackenbos is in Immigration and Naturalization Service Records, R.G. 85, 51993, National Archives, Washington, D.C.

30. Leroy Percy to J. B. Ray (former manager at Sunnyside), November 7, 1907, Percy Papers.

31. *The Outlook*, August 1907; and Theodore Roosevelt to Leroy Percy, August 13, 1907, R.G. 60, 100937.

32. Leroy Percy to President Theodore Roosevelt, November 13, 1907, R.G. 60, 100937, recounts Percy's entire argument; on Roosevelt's assurance of recalling Quackenbos, see Leroy Percy to John M. Parker, November 27, 1907, Percy Papers, and the attorney general to the president, November 26, 1907, R.G. 60, 100937.

33. Theodore Roosevelt to Albert Bushnell Hart, January 13, 1908, Albert Bushnell Hart Papers, Harvard University Archives, Cambridge, Mass.

34. Mary Grace Quackenbos to the attorney general, November 18, 1907, and Quackenbos to the attorney general, November 24, 1907, in R.G. 60, 100937.

35. The attorney general to Secretary of State Elihu Root, January 22, 1908, Department of State General Records, R.G. 59, microfilm 862, reel 539, National Archives, Washington, D.C.

36. Theodore Roosevelt to Charles Bonaparte, September 7, 1907, with clipping from the New York *Evening World*, "Woman Lawyer Aids Uncle Sam's Trust Busting," Charles Bonaparte Papers, Library of Congress, Washington, D.C.; attorney general to the president, September 10, 1907, and the president to the attorney general, September 19, 1907, R.G. 60, 100937. Of the last communication, the full letter is missing from both the Justice Department records and the Theodore Roosevelt Papers. Neither is there a copy in the Bonaparte Papers. The Justice Department records contain the

original document wrapper that summarizes the president's reply. See also Shofner, "Mary Grace Quackenbos: A Visitor Florida Did Not Want," 273–90, and Daniel, *Shadow of Slavery*, chapter 5.

37. Judge H. C. Niles, "To the Federal Grand Jury now in session at Vicksburg" and "The Reports of the Grand Jury" are misfiled in the Justice Department records with the Alonzo Bailey case, R.G. 60, 143691, National Archives, Washington, D.C.; Charles Wells Russell to the attorney general, reprinted in Russell, *Report on Peonage* (Washington, D.C., 1909).

38. Albert Bushnell Hart to Mr. President, January 10, 1908. Albert Bushnell Hart Papers, Harvard University Archives.

39. Leroy Percy to Gov. X. O. Pindall, December 14, 1907, copy in Albert Bushnell Hart Papers, Harvard University Archives.

40. J. E. Holland to Leroy Percy, October 11, 1907; and Percy to Holland, October 15, 1907, Percy Family Papers.

41. Albert Bushnell Hart to Mr. President, January 10, 1908, Albert Bushnell Hart Papers, Harvard University Archives; Leroy Percy to Mary Grace Quackenbos, October 19, 1907, file 100937, R.G. 60, National Archives; and John Gruenberg to the commissioner general [of Immigration], December 4, 1907, file 51993, R.G. 85, Immigration and Naturalization Service, National Archives.

42. Theodore Roosevelt to Albert Bushnell Hart, January 13, 1908, Hart Papers. On the interest in renewing the prosecution of Crittenden in Arkansas, see William G. Whipple, [U. S. attorney for the Eastern District of Arkansas] to the attorney general, January 24, 1908, file 100937, R.G. 60. On the satisfaction of the attorney general and his subordinates with the "moral" impact of the Crittenden prosecution, see Charles Wells Russell telegram to the attorney general, January 16, 1908, file 100937, R.G. 60; and also Russell, *Report on Peonage,* Government Printing Office, 1908, 14–16. On the fate of the alien contract labor law cases, see Charles Wells Russell to the secretary of Commerce and Labor, August 19, 1908; R. C. Lee to secretary of Commerce and Labor, December 7, 1908; and John Gruenberg to commissioner of Immigration, January 4, 1909, all in file 51993, R.G. 85.

43. *Congressional Record,* March 2, 1908, 2750–51 for Clark's remarks; 2747–50 for Humphries' remarks; 2746–47 for Williams' motion. For Representative Clark's original motion to create a special Congressional investigation committee, see *Congressional Record,* January 6, 1908, 482; and for John Sharp Williams' maneuvering on the Rules Committee, see *Congressional Record,* February 24, 1908, 2393–95; February 26, 2539; and February 28, 2688.

44. *United States Immigration Commission Reports,* volume II, 443–50.

45. This paragraph revises an earlier assesment I made of Federal manuscript census records for Sunnyside in 1910. See Randolph H. Boehm,

"Mary Grace Quackenbos and the Federal Campaign Against Peonage: The Case of Sunnyside Plantation," *Arkansas Historical Quarterly* (Spring 1991): 59. My error lay in the fact that by 1910, the plantation was divided into two enumeration districts, Lake Township and McConnell Township. In the 1900 census, the entire plantation was covered by McConnell Township, and I mistakenly used only the McConnell Township returns in making my earlier assessment of the 1910 population. The 1910 census shows 125 Italian families in McConnell Township and 37 in Lake Township. However, it may be erroneous to conclude that all 162 of these families lived on the Sunnyside Plantation in 1910. Although McConnell Township covered the northern two-thirds of the plantation, it extended northward beyond the plantation boundaries into Vaucluse, where Italians were known to be farming. Lake Township on the southern side of Sunnyside extended southward to include Red Leaf Plantation, where Italians were also known to be renters. How many of the 162 Italian families should be attributed to Red Leaf and Vaucluse is impossible to tell from the manuscript census.

I suggest "subtle coercion" as a possible means of keeping the renters on the land for several reasons, notably from Quackenbos's observations that the Italians felt themselves to be vulnerable to the fiat of the overseers. It also seems that Leroy Percy worked some sort of an arrangement with the Italian priest, Father Galloni, that made the cleric personally responsible for the debts of any Italians who left the property while in arrears. While the correspondence between Percy and the priest only hints at the agreement, Percy's implied threat would seem clear. The letter may bear quoting at some length:

> Dear Father,
>
> I am glad to hear that Mr Crittenden arranged matters in accordance with our conversation, in fact, even going beyond it, as he tells me he has written, or will write, Mr. Holland not to deduct the balances which we lost by Italians leaving us before Mr. Holland came on the place, although these balances were covered by your agreement.
>
> In the future, we will have to insist upon the agreement as made, namely, that where the Italians leave, the amount which we have advanced you, if they leave in debt, shall be refunded by you. I hope there will not be a great deal of this, if there is, it would be unfortunate for you and doubly disasterous to us . . .
>
> (Leroy Percy to Father Galloni,
> May 23, 1907, Percy Family Papers)

Nowhere in the documentary record is this "agreement" between the priest and the company spelled out. It may be that the priest took small advances to relieve distressed tenants himself and that he was only docked when those tenants left and not whenever any tenant left in debt. Yet, it is clear that the priest stood to suffer financially in at least some cases of abscondences. Under these circumstances, it seems reasonable that he encouraged distressed tenants not to abandon the plantation.

On the dissolution of Sunnyside, Santo Rossini, interview with the author, Lake Village, Ark., February 18, 1989. See also Brandfon, *Cotton Kingdom,* Ch. 6 and Baker, *Percys of Mississippi,* chapter 2.

46. Daniel, *Shadow of Slavery* shows conclusively the endurance of peonage in the South during the twentieth century. The massive peonage file of the Department of Justice in the National Archives provides a grim confirmation of Daniel's thesis. The victims, however, became almost exclusively African Americans.

Leroy Percy and Sunnyside

1. William Alexander Percy, *Lanterns on the Levee: Recollections of a Planter's Son* (1941; reprint, Baton Rouge: Louisiana State University Press, 1973); Lewis Baker, *The Percys of Mississippi: Politics and Literature in the New South* (Baton Rouge: Louisiana State University Press, 1983), 27–33.

2. Charles Scott to Stuyvesant Fish, December 26, 1904, Illinois Central Railroads Papers, Newberry Library, Chicago.

3. See Alfred H. Stone, "The Italian Cotton Grower: The Negro's Problem," *South Atlantic Quarterly* IV (January–October 1905): 42–47; see also Alfred H. Stone, "The Negro in the Yazoo-Mississippi Delta," *American Economic Association Publications,* 3rd ser., III (December 1901): 235–78.

4. Edmondo Mayor des Planches, *Attraverso gli Stati Uniti per L'Emigrazione Italiana* (Torino, Italy: Unione tipografica-editrice Toriuese 1913), 134–36, 217–18, 239–46.

5. See most especially, Mary Grace Quackenbos to Attorney General Charles J. Bonaparte, September 27 and October 25, 1907, Department of Justice Records, Record Group (hereinafter cited as R.G.) 60, 100937, National Archives, Washington, D.C.; and Quackenbos to Terence Powderly, Chief, Division of Information, Department of Commerce and Labor, September 28, 1907, R.G. 60, 100937. Vicksburg *Herald,* October 2, 1907.

6. See Mary Grace Quackenbos to Charles J. Bonaparte, September 28, 1907, R.G. 60, 100937.

7. Robert L. Brandfon, "The End of Immigration to the Cotton Fields," *Journal of American History* 50 (March 1964): 592 n.4, argues that proprietors renting land made a 6 to 7 percent return; sharecrop lands returned "sometimes more than 21 percent"; the average was 13.6 percent.

8. Leroy Percy to John Gruenberg, November 6, 1907, Records of the Department of Commerce and Labor, Immigration and Naturalization Service, Subject Correspondence, 1906–1932, R.G. 85, 51993, National

Archives, Washington, D.C.; Leroy Percy to James L. Watkins, February 20, 1908, Percy Family Papers, Mississippi Department of Archives and History, Jackson, Mississippi; hereinafter cited as Percy Papers.

9. Albert Bushnell Hart to President Theodore Roosevelt, January 10, 1908, in Benjamin Humphreys, March 2, 1908, *Congressional Record*, 60th Cong., 1st sess., 1908, 42, pt. 3, 2749.

10. See Mary Grace Quackenbos to Attorney General Bonaparte, September 28, 1907, R.G. 60, 100937.

11. Mary Grace Quackenbos to Attorney General Bonaparte, report, September 28, 1907 (forwarded to Secretary Elihu Root by Bonaparte, October 2, 1907, for transmission to the Italian ambassador), Department of State General Records, R.G. 59, 6912–47, National Archives, Washington, D.C.; hereinafter cited as Quackenbos Report, September 28, 1907.

12. See Louis Guida, "The Racconi-Fratesi Family: Italianatà in the Arkansas Delta," in Deirdre LaPin, Louis Guida, and Lois Pattillo, eds., *Hogs in the Bottom* (Little Rock: August House, 1982), 87 (quotation), and 86–100. See also Ernesto Milani, "Marchigiani and Veneti on Sunny Side Plantation," in Rudolph J. Vecoli, ed., *Italian Immigrants in Rural and Small Town America* (New York: The American Italian Historical Association, 1987), 18–30; and Amy Rose Scott, "Everyone Believed America Was a Place to Come: Sunnyside Plantation and Italian Immigrants," American Studies paper, fall 1989, University of North Carolina, kindly lent by the author.

13. For these figures, see George K. Holmes, *Supply of Farm Labor: Bulletin 94*, Bureau of Statistics, U.S. Department of Agriculture (Washington, D.C.: U.S. Government Printing Office, 1912), 39; the average farm laborer in Mississippi earned $12.92 per month or $155.04 per year, with board; without board, $18.56 per month and $222.72 per year.

14. See Leroy Percy to Attorney General Bonaparte, August 19, 1907, R.G. 60, 100937; Robert Foerster, *The Italian Emigration of Our Times* (1919; reprinted New York: Arno Press, 1969), 369.

15. John Savage to Leroy Percy, March 3, 1907, and Percy to Savage, March 9, 1907, Percy Papers; Leroy Percy to Mary Grace Quackenbos, August 17, 1907, R.G. 60, 100937.

16. "Materialmente stanno su per giù come in Italia, guadagnando di più" ("Materially they [the Sunnyside colonists] are better off than in Italy, earning more"), reported the Italian ambassador. Edmondo Mayor des Planches, "Nel Sud degli Stati Uniti," *Nuova Antologia*, 121, fasc. 820 (February 16, 1906), 18.

17. See Mary Grace Quackenbos to Attorney General Bonaparte, September 28, 1907, R.G. 60, 100937.

18. Cotton prices in the period were as follows: a disastrous 4.88 cents per pound, 1898–99; 1899–1900, 7.65 cents; 1900–01, 9.33 cents; 1901–02, 8.06

cents; 1902–03, 8.82 cents; 1903–04, 12.15 cents; 1904–05, 8.98 cents; 1905–06, 11.07 cents; 1906–07, 9.56 cents; 1907–08, 11.42 cents. The highest price before 1914 was 14.60 cents per pound in 1910–11. See James J. Lea, *Lea's Cotton Book Containing a Statistical History of the American Cotton Crop* (New Orleans: Press of Hauser Printing Co., 1914), 26.

19. See Mary Grace Quackenbos to Charles W. Russell, September 27, 1907, Department of Justice Records, R.G. 60, 74682, National Archives, Washington, D.C.

20. Mary Grace Quackenbos to Attorney General Bonaparte, July 20, 1907, R.G. 60, 74682, and Quackenbos to Bonaparte, August 1 (quotation), 15, 30, October 8, 1907, R.G. 60, 100937.

21. Humphreys, March 2, 1908, *Cong. Rec.*, 2747. Twenty-nine had escaped Sunnyside, leaving debts (see Quackenbos Report, September 28, 1907).

22. Quackenbos Report, September 28, 1907.

23. Vicksburg *Herald*, October 26, 1907, clipping, R.G. 60, 100937.

24. See Leroy Percy to President Theodore Roosevelt, November 14, 1907, R.G. 60, 100937.

25. Leroy Percy to Charles J. Bonaparte, August 19, 1907, R.G. 60, 100937.

26. Quoted in March 2, 1908, *Cong. Rec.*, 2749.

27. See John Gruenberg to F. P. Sargent, Report, December 10, 1907, in F. P. Sargent to Secretary of Commerce and Labor Oscar Straus, "Memorandum for the Secretary," December 18, 1907, R.G. 85, 51995. See also John Gruenberg to F. P. Sargent, November 4, 1907, R.G. 85, 51995.

28. On Vardaman's racism through Percy's eyes, see Leroy Percy to General Frank C. Armstrong, March 9, 1907, Percy Papers.

29. See Charles Scott to F. P. Sargent, December 26, 1904, Illinois Central Railroad Papers, Newberry Library, Chicago.

30. Quotation, Attorney General Bonaparte to Leroy Percy, August 23, 1907, R.G. 60, 100937; see also Humphreys, *Cong. Rec.*, 2747–48.

31. Leroy Percy to George Hebron, January 20, 1908, Percy Papers.

32. Leroy Percy to W. P. Dawson (editor, Chicago *Tribune*), December 21, 1907, Percy Papers; Leroy Percy to Attorney General Bonaparte, August 19, 1907, R.G. 60, 100937. Percy reported to Roosevelt that he asked the Labor Department to make inquiries and learned that "an investigation could be made by [the department] simply as to general conditions, not infractions of the law." Percy wanted Victor Clark's report published; it was apparently very favorable. See Leroy Percy to President Roosevelt, November 14, 1907, R.G. 60, 100937.

33. Some months before the firm was subjected to federal scrutiny, he had arranged with George Edgell, president of the Corbin Bank that held

the Sunnyside mortgage, to have the contracts translated into Italian. He argued, "It is just and because it deprives them of the plea that they did not understand what, in fact, they do understand." Leroy Percy to George Edgell, February 14, 1907, Percy Papers. One might have expected that after so many years in the business, a policy that could have been handled so easily would have been undertaken earlier. Percy clearly was open-handed; no fraud was intended.

34. Quackenbos Report, September 28, 1907.

35. Leroy Percy to Umberto Pierini, March 9, 1907, Percy Papers.

36. Leroy Percy to Edmondo Mayor des Planches, February 14, 1907, Percy Papers; Leroy Percy to Mary Grace Quackenbos, August 17, 1907, R.G. 60, 100937. (It should be added that the same apparently was not so true of Trail Lake where so many colonists left and about which the historian has very little information.)

37. Percy was not, however, unsolicitous about the Italians' health. See Leroy Percy to Dr. Dobson, June 1, 5, 1907, Percy Papers; Leroy Percy to Mary Grace Quackenbos, August 17, 1907, R.G. 60, 100937.

38. See John C. Willis, "Closing the Plantation Frontier: Small Farmers in the Yazoo-Mississippi Delta, 1865–1920," paper presented at the fifty-sixth annual meeting of the Southern Historical Association, New Orleans, November 1, 1990, kindly lent by the author.

39. Leroy Percy to John Gruenberg, November 6, 1907, R.G. 85, 51993.

40. See Leroy Percy to Will Dockery, March 8, 1907, and Percy to Father Galloni, May 23, 1907, Percy Papers; see also Rowland T. Berthoff, "Southern Attitudes Toward Immigration, 1865–1914," *Journal of Southern History* XVII (February 1951): 328–60; and, for a contrasting view, Brandfon, "End of Immigration to the Cotton Fields," 591–611.

41. See Leroy Percy to Britton and Koontz, December 29, 1904; and George Edgell to Leroy Percy, February 4, 1907, Percy Papers.

42. Leroy Percy to Mary Grace Quackenbos, August 17, 1907, R.G. 60, 100937.

43. Mary Grace Quackenbos to Attorney General Bonaparte, October 25, 1907, R.G. 60, 100937.

44. Leroy Percy to Charles J. Bonaparte, August 19, 1907, R.G. 60, 100937. See also Leroy Percy to Anna Clay Johnson, March 24, 1908, and Leroy Percy to Bolton Smith, October 8, 1907, Percy Papers.

45. Leroy Percy to John Savage, March 9, 1907, Percy Papers.

46. Mary Grace Quackenbos to Attorney General Bonaparte, August 14, 1907, R.G. 60, 100937.

47. Mary Grace Quackenbos to Leroy Percy, October 16, 1907, R.G. 60, 100937; Leroy Percy to Walker Percy, October 23, 1907, Leroy Percy to Lady

Percy McKinney (Mrs. C. J. McKinney), July 16, 1907, and Leroy Percy to Charles Scott, August 21, 1907, Percy Papers.

48. Jackson *Daily News*, October 26, 1907, clipping, R.G. 60, 100937.

49. See Leroy Percy to Walker Percy, November 18, 1907, Leroy Percy to J. S. McNeilly, November 19, 1907, and Leroy Percy to John Parker, November 27, 1907, Percy Papers. Mary Grace Quackenbos to Attorney General Bonaparte, September 28, 1907, R.G. 60, 100937.

50. Leroy Percy to John M. Parker, November 27, 1907, Percy Papers; see also Leroy Percy to President Theodore Roosevelt, November 14, 1907, and Attorney General Bonaparte to President Roosevelt, November 26, 1907, R.G. 60, 100937.

51. See William Loeb, Jr. (secretary to the president), to Attorney General Bonaparte, December 2, 1907, Special Correspondence with Theodore Roosevelt, Charles J. Bonaparte Papers, Library of Congress, Washington, D.C. On the Hart-Roosevelt correspondence, see Humphreys, March 8, 1907, *Cong. Rec.*, 2749; President Roosevelt to Albert Bushnell Hart, January 13, 1908, Albert Bushnell Hart Papers, Harvard University Archives, Cambridge, Mass., kindly supplied by Randolph H. Boehm.

52. William Loeb to Attorney General Bonaparte, December 2, 1907, Bonaparte Papers. Also, see President Roosevelt to Attorney General Bonaparte, September 7, 1907, Bonaparte Papers.

53. Leroy Percy to J. S. McNeilly, November 27, 1907, and Leroy Percy to J. B. Ray, November 7, 1907, Percy Papers. On Lee's conferences with Percy, see Leroy Percy to J. S. McNeilly, November 19, 20, December 31, 1907; Leroy Percy to John Parker, November 27, 1907; and Leroy Percy to [Senator] John Sharp Williams, November 30, 1907, Percy Papers. See also, Mary Grace Quackenbos to Charles Wells Russell, October 8, 1907, R.G. 60, 734682.

54. Percy quoted from Leroy Percy to Will Percy, January 16, 1908, and from Leroy Percy to Mrs. R. L. McLaurin, January 11, 1908, Percy Papers.

55. Leroy Percy to J. S. McNeilly, December 31, 1907; Leroy Percy to Will Percy, January 16, 1908; Leroy Percy to Judge Sidney M. Smith, February 14, 19, 21, 1908; Leroy Percy to Judge H. C. Niles, February 17, 1908, Percy Papers; Charles Wells Russell to Attorney General Bonaparte, January 16, 1908, telegram, R.G. 60, 100937. Quotation on Northumberland Percys in Merwyn E. James, *Society, Politics, and Culture: Studies in Early Modern England* (Cambridge: Cambridge University Press, 1986), 292.

56. Mary Grace Quackenbos to Charles Wells Russell, September 27, 1907, R.G. 60, 74862; Mary Grace Quackenbos to Attorney General Bonaparte, February 8, 1908, R.G. 60, 100937. See also Humphreys, March 8, 1908, *Cong. Rec.*, 2749, and Leroy Percy to J. M. F. Erwin, January 6, 1908, Percy Papers.

57. Pete Daniel, *The Shadow of Slavery: Peonage in the South, 1901–1969* (Urbana: University of Illinois Press, 1972).

58. See Quackenbos Report, September 28, 1907, and Mary Grace Quackenbos to Attorney General Bonaparte, October 8, 1907, R.G. 60, 73682.

59. See John Gruenberg to F. P. Sargent, November 4, 1907, R.G. 85, 51993, and Leroy Percy to F. P. Sargent, November 4, 1907, R.G. 85, 51993. On the role of the interpreter and Percy's displeasure, see Mary Grace Quackenbos to Attorney General Bonaparte, August 15, 1907, R.G. 60, 100937.

60. Leroy Percy to John Gruenberg, November 2, 1907, and Leroy Percy to F. P. Sargent, November 4, 1907, R.G. 85, 51993.

61. John Gruenberg to F. P. Sargent, November 4, 1907, R.G. 85, 51993. See also F. P. Sargent to Secretary of Commerce and Labor Straus, memorandum, December 18, 1907, R.G. 85, 51993.

62. F. P. Sargent to Oscar Straus, December 18, 1907, R.G. 85, 51993.

63. Mary Grace Quackenbos to Attorney General Bonaparte, July 20, 1907, R.G. 60, 73682.

64. Leroy Percy to John T. Savage, March 6, 1907, Percy Papers. See also Leroy Percy to George Edgell, March 25, 1907, Percy Papers.

65. See Leroy Percy to President Theodore Roosevelt, November 14, 1907, R.G. 60, 100937.

66. Leroy Percy to John Gruenberg, November 6, 1907, R.G. 85, 51993. He was also concerned that the Austro-Hungarian government had prohibited emigration to the American South. See Leroy Percy to J. S. McNeilly, November 19, 1907, Percy Papers. On his repudiation of the Italian authorities, see Leroy Percy to Vice Consul S. Moroni, May 30, 1908, Percy Papers.

67. Mary Grace Quackenbos to Attorney General Bonaparte, January 21, 1908, and Quackenbos to John M. Gracie, January 31, 1908, R.G. 60, 100937.

68. See Jay Tolson, *Pilgrim in the Ruins: A Life of Walker Percy* (New York: Simon & Schuster, 1992), 67–68.

69. Des Planches, "Attraverso gli Stati Uniti per L'Emigrazione Italiana," 142.

70. Des Planches, "Nel Sud degli Stati Uniti," 18.

71. See Leroy Percy to George Edgell, March 9, 1907, Percy Papers.

72. See Brandfon, "End of Immigration," 605–06.

73. Thomas H. Holloway, *Immigrants on the Land: Coffee and Society in São Paulo, 1886–1934* (Chapel Hill, N.C.: University of North Carolina Press, 1980), 139–74.

74. James R. Yerger to Leroy Percy, January 28, 1907, Percy Papers.

75. Walker Percy, *The Last Gentleman* (New York: Farrar, Straus and Giroux, 1968), 16.

Italian Migration: From Sunnyside to the World

1. A handy reference to the literature on Italian migration to the U.S. is contained in George E. Pozzetta, "Immigrants and Ethnics: The State of Italian American Historiography," *Journal of American Ethnic History* 9 (Fall 1990): 67–95.

2. Gianfausto Rosoli, *Un secolo di emigrazione italiana: 1876–1976* (Rome: Centro Studi Emigrazioné, 1978) is an excellent overview of this massive population movement.

3. The only volume yet available in English that has attempted to capture the extraordinary diversity of Italian migrations is Robert Franz Foerster's classic work, *The Italian Emigration of Our Times* (Cambridge, Mass.: Harvard University Press, 1919), which ironically was written in the midst of the great Italian mass migration to the United States. See also George E. Pozzetta and Bruno Ramirez, eds., *The Italian Diaspora: Migration Across the Globe* (Toronto: Multicultural History Society of Ontario, 1992), which attempts to refocus contemporary research to a global agenda.

4. A theoretical introduction to the broader patterns of movement is contained in Dirk Hoerder, "International Labor Markets and Community Building by Migrant Workers in the Atlantic Economies," in Rudolph J. Vecoli and Suzanne Sinke, eds., *A Century of European Migrations, 1830–1930* (Urbana: University of Illinois Press, 1992), 78–110. For more information on Italian migration, see Franco Ramella, "Emigration from an Area of Intense Industrial Development: The Case of Northwestern Italy," in Vecoli, *A Century*, 261–76; and Donna Gabaccia, "International Approaches to Italian Labour Migration," in Pozzetta and Ramirez, *The Italian Diaspora*, 21–36.

5. John and Leatrice MacDonald, "Chain Migration, Ethnic Neighborhood Formation and Social Networks," *Milbank Memorial Fund Quarterly* 42 (January 1964): 82–91; "Urbanization, Ethnic Group and Social Segmentation," *Social Research* 29 (Winter 1962): 443–48; "Italy's Rural Social Structure and Emigration," *Occidente* 12 (1956): 437–56. For an outstanding case study of centuries-old migration traditions built upon craft skills and kin networks, see Patrizia Audenino, "The Paths of the Trade: Italian Stonemasons in the United States," *International Migration Review* 20 (1986): 779–95.

6. From among the dozens of community studies that have documented the Italian immigrant experience, see Gary R. Mormino, *Immigrants on the Hill: Italian Americans in St. Louis, 1882–1982* (Urbana: University of Illinois Press, 1986); and William M. DeMarco, *Ethnics and Enclaves: Boston's Italian North End* (Ann Arbor: University Microfilms, 1981).

7. Robert Harney, "The Commerce of Migration," *Canadian Ethnic Studies* 9 (1977), 42–54; "From Central Asia to the Soo: The Adventure of Italian Migration," *Mosaico*, 1 (March 1976): 18–21.

8. John E. Zucchi, "Occupations, Enterprise, and the Migration Chain: The Fruit Traders from Termini Imerese in Toronto, 1900–1930," *Studi Emigrazione* 77 (Marzo 1985): 68–79; John E. Zucchi, *The Little Slaves of the Harp: Italian Street Musicians in Nineteenth Century Paris, London, and New York* (Montreal: McGill University Press, 1992). Zucchi's *Italians in Toronto: Development of a National Identity, 1875–1935* (Montreal: McGill University Press, 1988), places these sorts of movements in the context of community growth and identity formation.

9. Ernesto Milani, "Peonage at Sunnyside and the Reaction of the Italian Government," *Arkansas Historical Quarterly* 50 (Spring 1991): 36–37. See also Milani's essay in this volume.

10. Humbert Nelli, "The Italian Padrone System in the United States," *Labor History* 2 (Spring 1964): 153–67; Luciano J. Iorizzo, "The Padrone and Immigrant Distribution," in *The Italian Experience in the United States*, Silvano M. Tomasi and Madeline H. Engel, eds., (New York: Center for Migration Studies, 1970), 43–75.

11. Robert Harney, "A Case Study of Padronism: Montreal's Kind of Italian Labor," *Labour/Le Travailleur* 4 (1979): 57–84; "The Padrone and the Immigrant," *Canadian Review of American Studies* 5 (Fall 1974): 101–18; "The Padrone System and Sojourners in the Canadian North, 1885–1920," in George E. Pozzetta, ed., *Pane e Lavoro: The Italian Working Class* (Toronto: Multicultural History Society of Ontario, 1980), 119–37.

12. Richard Gambino, *Vendetta: A True Story of the Worst Lynching in America, the Mass Murder of Italian-Americans in New Orleans in 1891, the Vicious Motivations Behind It, and the Tragic Repercussions that Linger to This Day* (New York: Doubleday, 1977). See also Humbert Nelli, "The Hennessy Murder and the Mafia in New Orleans," *Italian Quarterly* 19 (1975): 77–95.

13. John V. Baiamonte, Jr., *Spirit of Vengeance: Nativism and Louisiana Justice, 1921–1924* (Baton Rouge: Louisiana State University Press, 1986).

14. George E. Pozzetta, "Gino C. Speranza: Reform and the Immigrant," in Pozzetta and David R. Colburn, eds., *Reform and Reformers in the Progressive Era* (Westport, Conn.: Greenwood Press, 1983), 47–70. For a sense of contemporary reporting, see Gino C. Speranza, "Labor Abuses Among Italians," *Charities* 12 (1904): 448–49; "The Italian Foreman as a Social Agent," *Charities* 11 (1903): 27; "Force Labor in West Virginia," *Outlook* 74 (1903): 407–10; "Italian Farmers in the South," *Charities* 15 (1906): 307–08.

15. Luciano J. Iorizzo and Salvatore Mondello, *The Italian Americans*, 2d ed. (Boston: Twayne Publishers, 1980), 288–89, has a chart that lists fifty-six Italian rural communities in the United States, 1900–10, located in twenty different states. Foerster's classic work provides a global overview of agricultural colonization projects, but it also discusses various experiments in, among other locations, Fredonia, New York; Valdese, North Carolina; Vineland, New Jersey; Bryan, Texas; and Cumberland, Wisconsin.

16. Iorizzo and Mondello, *The Italian Americans*, 139–43. See also John L. Mathews, "Tontitown, A Story of Conservation of Men," *Everybody's Magazine* 20 (January 1909): 3–13, and John Andreozzi, "Italian Families in Cumberland [Wisconsin]," in Rudolph J. Vecoli, ed., *Italian Immigrants in Rural and Small Town America* (New York: Center for Migration Studies, 1987): 110–25, for a study of the 4th largest Italian rural settlement.

17. Foerster, *The Italian Immigration*, 369.

18. Vecoli, *Italian Immigrants*.

19. Rudolph J. Vecoli, "The Italians," in *They Chose Minnesota: A Survey of the State's Ethnic Groups*, June Drenning Holmquist, ed. (St. Paul: Minnesota Historical Society, 1981), 449–71. See also his *The People of New Jersey* (Princeton: Van Nostrand Publishers, 1965) for coverage of rural settlements in New Jersey.

20. Valentine J. Belfiglio, *The Italian Experience in Texas* (Austin: Eakin Publishers, 1983); and "Italians in Rural and Small Town Texas," in Vecoli, *Italian Immigrants*, 31–49.

21. J. Vencenza Scarpaci, *Italian Immigrants in Louisiana's Sugar Parishes: Recruitment, Labor Conditions, and Community Relations* (New York: Arno Press, 1980).

22. Andrew F. Rolle, *The Immigrant Upraised: Italian Immigrants and Colonists in an Expanding America* (Norman: University of Oklahoma Press, 1968).

23. George E. Pozzetta, "The Italians of New York City, 1890–1914," Ph.D. diss., University of North Carolina, 1971, discusses a number of initiatives designed to resettle urban-dwelling Italians in rural locations. See also Foerster, 369–73.

24. Interestingly, the social mobility patterns of immigrants in rural locations has received almost no attention. For works exploring the nature of urban mobility see Thomas Kessner, *The Golden Door: Italian and Jewish Immigrant Mobility in New York City* (New York: Oxford University Press, 1977); and Stephan Thernstrom, *The Other Bostonians: Poverty and Progress in the American Metropolis* (Cambridge, Mass.: Harvard University Press, 1973).

25. Betty Boyd Caroli, *Italian Repatriation from the United States, 1900–1914* (New York: Center for Migration Studies, 1973).

26. Francesco Cerase, "A Study of Italian Migrants Returning from the U.S.A.," *International Migration Review* 1 (1967): 67–74; "Expectations and Reality: A Case Study of Return Migration from the United States to Southern Italy," *International Migration Review* (1974): 245–63.

27. Dino Cinel, "The Seasonal Emigration of Italians in the Nineteenth Century," *Journal of Ethnic Studies* 10 (Spring 1982), 43–68.

28. The best guide to this enormous literature is Silvano M. Tomasi and Edward Stibili, eds., *Italian Immigrants and the Catholic Church: An*

Annotated Bibliography, 2d ed. (New York: Center for Migration Studies, 1992).

29.   See, for example, Silvano M. Tomasi, *Piety and Power: The Role of Italian Parishes in the New York Metropolitan Area* (New York: Center for Migration Studies, 1975); Gary R. Mormino, "The Church Upon the Hill: Italian Immigrants in St. Louis, Missouri," *Studi Emigrazione* 19 (1982): 203–24.

30.   P. W. Bardaglio, "Italian Immigrants and the Catholic Church in Providence, R.I.," *Rhode Island History* 34 (May 1975): 47–57.

31.   Richard Varbero, "Philadelphia's South Italians and the Irish Church: A History of Cultural Conflict," in *The Religious Experience of Italian Americans,* Silvano M. Tomasi, ed. (New York: Center for Migration Studies, 1974), 33–54.

32.   Pietro diDonato, *Immigrant Saint: The Life of Mother Cabrini* (New York: McGraw-Hill, 1960); Serio C. Lorit, *Frances Cabrini* (New York: New City Press, 1970); Marco Caliaro and Mario Francesconi, *John Baptist Scalabrini: Apostle to Emigrants,* trans. Alba I. Zizzamia (New York: Center for Migration Studies, 1977); Edward Stibili, "The Interest of Bishop Giovanni Battista Scalabrini of Piacenza in the 'Italian Problem'," in *The Religious Experience of Italian Americans,* 11–30.

33.   Rudolph J. Vecoli, "Prelates and Peasants: Italian Immigrants and the Catholic Church," *Journal of Social History* 2 (Spring 1969): 217–68.

34.   Gary R. Mormino and George E. Pozzetta, *The Immigrant World of Ybor City: Italians and Their Latin Neighbors in Tampa, 1885–1985* (Urbana: University of Illinois Press, 1987), 210–32.

35.   Joseph Velikonja, "The Italian American Periodical Press, 1836–1980," in Lydio F. Tomasi, ed., *Italian Americans: New Perspectives* (New York: Center for Migration Studies, 1984), 248–59. Pietro Russo's forthcoming *Italian American Periodical Press, 1836–1980: Survey and Catalogue,* has already identified more than 2,400 separate titles which have appeared in the United States.

36.   See Gianfausto Rosoli, "From the Inside: Popular Autobiography by Italian Immigrants in Canada," in Pozzetta and Ramirez, *The Italian Diaspora,* 175–92, for a discussion of the utility of this kind of source.

37.   Rudolph J. Vecoli, ed., *Rosa: The Life of an Immigrant Woman* (Minneapolis, 1970).

38.   See Salvatore John LaGumina, *The Immigrants Speak: Italian Americans Tell Their Story* (New York: Center for Migration Studies, 1979); Michael LaSorte, *La Merica: Images of Italian Greenhorn Experience* (Philadelphia: Temple University Press, 1985).

39.   Jerre Gerlando Mangione's classic, *Mount Allegro: A Memoir of Italian-American Life* (Boston: Houghton Mifflin, 1943), remains extremely

valuable, as is Pietro diDonato's evocative *Christ in Concrete* (New York: Bobbs-Merrill, 1937). For more recent efforts, see Joseph Napoli, *A Dying Cadence: Memories of a Sicilian Childhood* (Baltimore: Privately Published, 1986) and Vincent Panella, *The Other Side: Growing Up Italian in America* (Garden City: Doubleday Press, 1979).

Index

Hawkins, Hamilton R., 22
Hebron Plantation, 10, 22
Higgs, Robert, 31
Humphreys, Congressman Benjamin G.,
  72, 83–85
Hyner Plantation, 10, 22

Illinois Central Railroad, 78
Immigrant press, 101
Immigration Bureau, 18
Immigration Commission, 109
Iorizzo, Luciano J., 97
Irish Catholics, 101
Italian Ambassador, (see Italian
  Government)
Italian Government: actions constrained,
  47; attitude toward plight of Italian
  immigrants, 39; ambassador com-
  plains, 49; ambassador requests
  investigation, 62; bureaucracy, 47;
  Des Planches visits Sunnyside, 45–46,
  78; diplomats, 18, 48; embassy, 41–43,
  48; enmity of toward Sunnyside
  experiment, 24; Fava (Ambassador
  Francesco Saverio), 40; inquiries of,
  58; Italian Royal Ministry of Foreign
  Affairs, 41; loses interest, 44; monitors
  Quackenbos's efforts, 75; Oldrini's
  interaction with, 41; orders investigation, 43; Percy queries ambassador,
  86; prohibits recruitment to southern
  states, 47, 91–92; Quackenbos's report
  influences, 92; Quackenbos sends
  report to, 66; receives complaints, 51;
  (Don Emanuele) Ruspoli's role, 18,
  20, 41, 44
Italian Royal Ministry of Foreign
  Affairs, 41
Italian–American labor relations, 39
Italians: advantages to at Sunnyside, 81;
  agents lure immigrants, 44; in
  Argentina, 48; and arrival of second
  group at Sunnyside, 21; and artesian
  wells, 87; in the Balkins, 96; and
  (Father) Bandini, 41–43; and blacks,
  22, 50; in Brazil, 93, 99; and Bureau of

Information and Protection for
  Italian Emigration, 40; in Canadian
  Northwest, 96; Catholics, 101;
  cheating of, 79; as less than full
  citizens, 87; complaints of, 43, 46, 49,
  51; committee formed to present, 42;
  conditions, 47, 83; and contract labor
  laws, 46; contract with Corbin, 18–19;
  contracts, 45; Corbin, scheme, 40;
  Corbin's experiment with Italian
  labor, 78; Crittenden Company,
  53–54, 78; in Cuba, 96; curtailing of
  Italian immigration to the lower
  Miss Valley, 75, 78; decline of population of Italians at Sunnyside, 22;
  Giovanni D'Elpidio heads committee to present complaints, 42;
  disillusionment among, 50; George
  Edgell agrees to demands, 44; effect
  of Italian's leaving, 92; enticement
  with false promises, 79; end of experiment, 47; failure of experiment, 24;
  failure to offer Sunnyside land for
  sale to, 93; first colonizing effort,
  49–50; flee plantation, 85; in France,
  96; on Gracie (John M.) plantation,
  20; hostility toward, 20; Immigrant
  agency in New York, 18; improvements needed, 84; intimidation of,
  44, 63; kidnapped, 60–61, 63, 89; as a
  labor experiment with, 3, 17–18, 36,
  39; and labor experience of, 26; labor
  needs served, 49; labor problems of
  the South solved by, 21; in Louisiana
  sugar fields, 99; lynching of in New
  Orleans, 20; mail rifled, 90; malcontents among, 51; migration to
  other areas, 96, 99; in Moscow, 96;
  northern rather than southern
  Italians desired, 20, 35; number of
  families at Sunnyside, 52, 69, 74;
  number of families in Delta, 47, 51;
  other southern states try to get
  Percy's Italians (he thinks), 91; in
  Paris, 96; Percy's attitude toward, 88,
  92; in Peru, 96; problems at Sunny-